SpringerBriefs in History of Science and Technology

Series Editors

Gerard Alberts, University of Amsterdam, Amsterdam, The Netherlands

Theodore Arabatzis, University of Athens, Athens, Greece

Bretislav Friedrich, Fritz Haber Institut der Max Planck Gesellschaft, Berlin, Germany

Ulf Hashagen, Deutsches Museum, Munich, Germany

Dieter Hoffmann, Max-Planck-Institute for the History of Science, Berlin, Germany

Simon Mitton, University of Cambridge, Cambridge, UK

David Pantalony, Ingenium - Canada's Museums of Science and Innovation / University of Ottawa, Ottawa, Canada

Matteo Valleriani, Max-Planck-Institute for the History of Science, Berlin, Germany

The *SpringerBriefs in the History of Science and Technology* series addresses, in the broadest sense, the history of man's empirical and theoretical understanding of Nature and Technology, and the processes and people involved in acquiring this understanding. The series provides a forum for shorter works that escape the traditional book model. SpringerBriefs are typically between 50 and 125 pages in length (max. ca. 50.000 words); between the limit of a journal review article and a conventional book.

Authored by science and technology historians and scientists across physics, chemistry, biology, medicine, mathematics, astronomy, technology and related disciplines, the volumes will comprise:

1. Accounts of the development of scientific ideas at any pertinent stage in history: from the earliest observations of Babylonian Astronomers, through the abstract and practical advances of Classical Antiquity, the scientific revolution of the Age of Reason, to the fast-moving progress seen in modern R&D;
2. Biographies, full or partial, of key thinkers and science and technology pioneers;
3. Historical documents such as letters, manuscripts, or reports, together with annotation and analysis;
4. Works addressing social aspects of science and technology history (the role of institutes and societies, the interaction of science and politics, historical and political epistemology);
5. Works in the emerging field of computational history.

The series is aimed at a wide audience of academic scientists and historians, but many of the volumes will also appeal to general readers interested in the evolution of scientific ideas, in the relation between science and technology, and in the role technology shaped our world.

All proposals will be considered.

Bruce Cameron Reed

The Frisch-Peierls Memorandum

The Founding Document of the Nuclear Age

 Springer

Bruce Cameron Reed
Alma College
Alma, MI, USA

ISSN 2211-4564 ISSN 2211-4572 (electronic)
SpringerBriefs in History of Science and Technology
ISBN 978-3-031-95928-8 ISBN 978-3-031-95929-5 (eBook)
https://doi.org/10.1007/978-3-031-95929-5

© The Editor(s) (if applicable) and The Author(s), under exclusive license to Springer Nature Switzerland AG 2025

This work is subject to copyright. All rights are solely and exclusively licensed by the Publisher, whether the whole or part of the material is concerned, specifically the rights of translation, reprinting, reuse of illustrations, recitation, broadcasting, reproduction on microfilms or in any other physical way, and transmission or information storage and retrieval, electronic adaptation, computer software, or by similar or dissimilar methodology now known or hereafter developed.
The use of general descriptive names, registered names, trademarks, service marks, etc. in this publication does not imply, even in the absence of a specific statement, that such names are exempt from the relevant protective laws and regulations and therefore free for general use.
The publisher, the authors and the editors are safe to assume that the advice and information in this book are believed to be true and accurate at the date of publication. Neither the publisher nor the authors or the editors give a warranty, expressed or implied, with respect to the material contained herein or for any errors or omissions that may have been made. The publisher remains neutral with regard to jurisdictional claims in published maps and institutional affiliations.

This Springer imprint is published by the registered company Springer Nature Switzerland AG
The registered company address is: Gewerbestrasse 11, 6330 Cham, Switzerland

If disposing of this product, please recycle the paper.

Preface

The Magna Carta. The Gutenberg Bible. The United States Declaration of Independence. Readers familiar with Western history will know that documents such as these are considered foundational to modern civilization. They are milestones in human development that had immediate impact and continue to be relevant today.

Lists of pivotal historical documents rarely contain material of a scientific nature. But there is one twentieth-century technical document that arguably should appear in every such compilation: The Frisch-Peierls memorandum of 1940.

In brief, the Frisch-Peierls (FP) memorandum was the first document to reach high government levels that laid out the technical basis for how to construct nuclear weapons, what effects could be expected from such weapons, and their military, strategic, and ethical implications. This astonishingly perceptive manuscript initiated the British government's World War II nuclear program, which in 1943 merged with the larger United States Manhattan Project. The Frisch-Peierls memorandum is truly a foundational document of the nuclear age. Physicist and science writer Jeremy Bernstein has called it a memorandum that changed the world.

Who were Frisch and Peierls? How did they come to prepare their memorandum? What were the details of its contents? How did they get it to responsible authorities during the war? How accurate were their predictions? What became of them? What were their thoughts on the nuclear age they helped birth?

This book analyses the FP memorandum, including its technical background, qualitative and scientific contents, and its influence on the wartime Allied nuclear program. I also offer biographical sketches of the lives and careers of Frisch and Peierls.

As described below, several papers analyzing various aspects of the memorandum have been published in specialized technical and history-of-science journals. My motivation in preparing this book is to bring appreciation and understanding of the FP memorandum to a wider audience, particularly present-day students interested in the physics of nuclear weapons and nuclear policy/strategy. My primary audience is intended to be physicists, but I hope that the narrative material will be of interest to readers interested in broadening their command of the history of nuclear issues and how we came to today's nuclear world.

Fig. 1 Left: Otto Frisch (1904–1979) Los Alamos identity badge photo. *Source* https://commons.wikimedia.org/wiki/ File: Otto Frisch Los Alamos ID badge photo.jpg. Credit statement: Unless otherwise indicated, this information has been authored by an employee or employees of the Los Alamos National Security, LLC (LANS), operator of the Los Alamos National Laboratory under Contract No. DE-AC52-06NA25396 with the U.S. Department of Energy. The U.S. Government has rights to use, reproduce, and distribute this information. The public may copy and use this information without charge, provided that this Notice and any statement of authorship are reproduced on all copies. Neither the Government nor LANS makes any warranty, express or implied, or assumes any liability or responsibility for the use of this information. Right: Rudolf Peierls (1907–1995). Source Public domain, https://commons.wikimedia.org/wiki/ File: Peierls, Rudolf 1966 Göttingen.jpg

As with many historically important documents, the FP memorandum is best appreciated in retrospect. To this end, I hope that readers bring some knowledge of the history and physics of the Manhattan Project, which is well-covered in much existing literature. Some of the background physics is surveyed in Chap. 1, but in general I will go into details only where they are pertinent to some aspect of the memorandum; this is a monograph, not a history of nuclear physics. But I do hope you know what alpha and beta decay are, what an isotope is, what a notation like $^1_0 n$ or $^{235}_{92} U$ means, what is meant by a half life, a mean free path, a reaction cross-section, and a MeV. If these are familiar, you are off to a good start.

Some further information and orientation. The memorandum was written in March 1940 by Otto Frisch and Rudolf Peierls, refugee European physicists then at the University of Birmingham in Britain; see Fig. 1. Their document described how uranium-235 could be used to set up a chain reaction to create an extremely powerful bomb, how such bombs would function, and what would be their effects. Their manuscript reached the UK government's Committee on the Scientific Survey of Air Warfare, which was headed by Sir Henry Tizard. This prompted the formation of the wartime British nuclear program. Among other native and naturalized British scientists, Frisch and Peierls subsequently moved to the United States to participate in the work that culminated with the July 1945 Trinity test that set the stage for the bombings of Hiroshima and Nagasaki.

The memorandum comprised two parts, a short qualitative document intended for government officials, plus a longer technical appendix, the beginning of which is shown in Fig. 2. The former was found among the papers of Sir Henry Tizard by historian Ronald Clark, and appears in his 1965 biography of Tizard. Somewhere

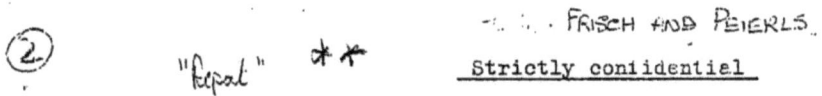

Fig. 2 The top of the first page of the technical memorandum. Courtesy Bodleian library, Oxford University

along the line, the two parts became separated, and the technical part is now held by the Bodleian Library of Oxford University. This has been reprinted in several sources, but, unfortunately, these often contain various errors. This likely began with Margaret Gowing's excellent history of the British nuclear program, *Britain and Atomic Energy 1939–1945* (Gowing, 1964), where typographical errors crept into a key paragraph and a formula was printed in a potentially ambiguous way. This was reprinted without correction in a paper by British nuclear historian Lorna Arnold (Arnold, 2003), in Robert Serber's *Los Alamos Primer* (Serber, 1992), in Ferenc Szasz's book on British scientists at Los Alamos (Szasz, 1992), and in a posthumous memoir (Peierls, 1997). The correct version does appear in Peierls' collected papers (Dalitz and Peierls, 1997), but this can be difficult to find. The present book includes a correct re-typing of the memorandum accompanied by numbered paragraphs to facilitate discussion in the main text; I obtained a copy of the original technical part of the memorandum from the Bodleian Library.

Even with a correct version in hand and armed with a reasonable knowledge of the background physics, a reader who comes to the memorandum cold will find it very difficult reading. It was written for an audience of government officials; Frisch and Peierls included no references for their adopted parameter values or details as to how they arrived at derivations for quantities such as the critical mass for and explosive power of a fission weapon. Compounding this is that they underestimated the critical mass of uranium-235 by a factor of about 80. This was not a fundamental error; their analysis was in fact quite sound. The underestimate traced to an overoptimistic prediction of the fission properties of ^{235}U, which they had to estimate because various quantities had not yet been measured. But for a newcomer, this honest scientific "error" certainly adds to the confusion.

Why do I believe that now is a good time for a synoptic survey of the memorandum? Over the last several years, various re-analyses of Frisch and Peierls' physics have developed insights as to what contemporary information might have guided their thinking. These include Bernstein (2011), Chadwick (2021), Lestone et al. (2021), McCauley (2025), Pearson (2024), Pearson and Reed (2024), Reed (2022), Reed (2024), among others. As a result of these efforts, it is fair to say that we now have some reasonable speculations as to what Frisch and Peierls might have done. We will of course never know for certain what they were thinking, but this book surveys both the fundamental physics involved and the strengths and weaknesses of these various analyses. In the end, I hope that readers take away a thorough understanding of a fascinating document.

This book comprises five chapters and seven appendices. Chapter 1 offers a brief review of nuclear research from early 1934 onward that led, by early 1940, to the discovery, verification, and interpretation of nuclear fission. By that time, experimental and theoretical work had firmly established uranium-235 as the "explosive" fissile isotope of that element. Chapter 2 turns the clock back to give biographical sketches of the lives and careers of Frisch and Peierls, how they came to both be in Birmingham in early 1940 following the discovery of fission, how the memorandum came to be prepared, its journey through the British government, and their transfer to the United States. This chapter also covers the memorandum's role in stimulating the wartime British and American nuclear programs and the sometimes fraught relationship between them. Chapter 3 presents an analysis of the qualitative part of the memorandum (reproduced in Appendix A), with an emphasis on Frisch and Peierls' anticipation of strategic, ethical, arms-race, cold-war strategies, and civil-defense issues associated with nuclear weapons.

Chapter 4 is the heart of this work: A detailed paragraph-by-paragraph deconstruction of the technical content of the memorandum (reproduced in Appendix B), focusing on the underlying scientific concepts, mathematical analyses, errors in various reprintings, and what contemporary information might have borne on Frisch and Peierls' choices of parameter values and numerical estimates. I also review what numbers they would have calculated had they used other arguably applicable and present-day values of various nuclear parameters. Supporting derivations appear in Appendices C and D for readers who wish to dig deeper into the physics. Chapter 5 offers a few closing remarks. Appendix E offers a detailed proof of the equivalence of two methods of computing the critical radius of a spherical core of fissile material in the limiting case where the radius is "large". Appendices F and G offer a glossary of mathematical symbols used throughout the text and a table of physical constants.

To close this Introduction, a brief comment on another nuclear document. Readers familiar with the Manhattan Project will know that in August, 1939, Albert Einstein signed a letter prepared by Leo Szilard and Eugene Wigner warning of the possibility of nuclear weapons that made its way to President Franklin Roosevelt in October of that year. This document initiated the American nuclear program with its enormous funding and facilities. Frisch and Peierls and the US-based group knew nothing of each other's documents until much later. Is not the Einstein letter just as foundational as the memorandum? I focus on the memorandum because it explores the technical details of a bomb, giving quantitative estimates of critical mass, destructive power, demands of isotope separation requirements, and radioactive effects; the Einstein letter dealt with none of these issues. Given the subsequent history of the American nuclear program, perhaps it is a tie, but for physicists the memorandum is by far the more fascinating document.

Alma, USA   Bruce Cameron Reed

References

Arnold, L. (2003). The history of nuclear weapons: The Frisch-Peierls memorandum on the possible construction of atomic bombs of February 1940. *Cold War History, 3*, 111–126.
Bernstein, J. (2011). A memorandum that changed the world. *American Journal of Physics, 79*(5), 440–446
Chadwick, M. (2021). Nuclear science for the Manhattan Project and comparison to today's ENDF data. *Nuclear Technology, 207*(S1), S24–S61
Dalitz, R. H., Peierls, R. E. (1997). *Selected scientific papers of Sir Rudolf Peierls with commentary.* World Scientific.
Gowing, M. (1964). *Britain and atomic energy 1939–1945.* St. Martin's Press.
Lestone, J. P., Rosen, M. D., & Adsley, P. (2021). Comparison between historic nuclear explosion yield formulas. *Nuclear Technology, 207*(S1), S352–S355
McCauley, J. (2025). Predictions of critical radii for reactors and bombs 1939–45 including the Frisch-Peierls memorandum. *The European Physical Journal H, 50*(1)
Pearson, J. M. (2024). Comments on the Frisch-Peierls estimate of the critical mass of a uranium fission bomb. *Nuclear Technology, 210*(6), 1078–1082
Pearson, J. M., & Reed, B. C. (2024). Remarks on the yield of fission bombs. *American Journal of Physics, 92*(9), 680–685
Peierls, R. (ed.) (1997). *Atomic histories.* American Institute of Physics.
Reed, B. C. (2022). Comments on the physics of the Frisch-Peierls memorandum. *Nuclear Technology, 208*(12), 1890–1893
Reed, B. C. (2024). Revisiting the Frisch-Peierls memorandum. *The European Physical Journal H, 49*, 49(6)
Serber, R. (1992). *The Los Alamos primer.* University of California Press.
Szasz, F. M. (1992). *British scientists and the Manhattan Project: The Los Alamos years.* St. Martin's Press.

Acknowledgements Over some three decades, I have benefited from conversations, correspondence, suggestions, well-deserved criticism, willingness to read endless drafts, and general encouragement from John Abelson, Joseph-James Ahern, John Altholz, Dana Aspinall, Michael Atlas, Karen Ball, Albert Bartlett, Jeremy Bernstein, Dick Bowker, Peter Burns, Alan Carr, David Cassidy, Mark Chadwick, John Coster-Mullen, Steve Croft, Peter Dawson, Gene Deci, Michael DeRobertis, Carleen Dewit, Cassiano Endre de Oliveira, Eric Erpelding, Patricia Ezzell, Charles Ferguson, Miriam Focaccia, Henry Frisch, Patrick Furlong, Ed Gerjuoy, John Gibson, Chris Gould, Dick Groves, Robert Hayward, Dave Hafemeister, Miriam Hiebert, Lorraine Hill, Art Hobson, Dieter Hoffmann, Steuard Jensen, Lisa Jylänne, Annelies Kersbergen, Patricia Kinnee, Tim Koeth, Vern Koslowsky, Gilles Labrie, William Lanouette, Irving Lerch, John Lestone, Harry Lustig, Mike Magras, Jeffrey Marque, Albert Menard, Tony Murphy, Lorne Nelson, Robert S. Norris, Steve Olson, John Palimaka, Mark Paris, Mike Pearson, Peter Pesic, Patrizia Piredda, Manfred Popp, Klaus Rohe, Bob Sadlowe, John Schreiner, Tom Semkow, Frank Settle, Ruth Sime, D. Ray Smith, Whitney Spivey, Ute Stargardt, Roger Stuewer, Arthur Tassel, Linda Thomas, Michael Traynor, George Wagner, Alex Wellerstein, Bill Wilcox, John Yates, and Pete Zimmerman. John Altholz in particular has been relentless in drawing typos and clunky grammar to my attention. If I have forgotten you, know that you are in this list in spirit. A few of these individuals are, sadly, no longer with us but are fondly remembered. Angela Lahee and her colleagues at Springer once again deserve a tip of my hat for supporting this project. In addition, the fingerprints of a lifetime's worth of family members, teachers, classmates, professors, mentors, colleagues, students, collaborators, and friends are all over these pages; a work like this is never accomplished in isolation.

But most of all again is Laurie.

Competing Interests The author has no competing interests to declare that are relevant to the content of this manuscript.

Contents

1	**Nuclear Fission: A Review**	1
	1.1 Neutrons, Induced Radioactivity, and Transuranic Elements	1
	1.2 Discovery and Verification of Fission	6
	1.3 Nuclear Parity and Neutrons Fast and Slow	12
	1.4 Bohr and Wheeler: The Fission Barrier and Chain Reactions	16
	1.5 Criticality	20
	1.6 Bohr Verified	21
	References	22
2	**Frisch and Peierls**	25
	2.1 Otto Robert Frisch	25
	2.2 Rudolf Ernst Peierls	33
	2.3 MAUD, Roosevelt, and Churchill: Bomb Politics	41
	References	48
3	**The Memorandum: Qualitative Part**	51
	References	56
4	**The Memorandum: Technical Part**	57
	4.1 Paragraphs [1] and [2]: Uranium and Neutron Bombardment	58
	4.2 Paragraphs [3]–[5]: Chain Reactions	58
	4.3 Paragraph [6]: No Slow-Neutron Bomb	60
	4.4 Paragraphs [7]–[9]: Possibility of a Fast-Neutron Bomb	65
	4.5 Paragraph [10]: The Critical Radius and Mass	66
	4.5.1 The Critical Radius	66
	4.5.2 The Fission Cross Section	67
	4.6 Paragraph [11]: Speed of the Reaction	70
	4.7 Paragraph [12]: The Yield	73
	4.8 Paragraphs [13]–[15]: Triggering and Predetonation	75
	4.9 Paragraphs [16]–[18]: Thermal Diffusion and Isotope Separation	77
	4.10 Paragraphs [19]–[24]: Radiation Effects	79
	References	82

Epilogue 85
Appendix A: The Memorandum: Qualitative Part 87
Appendix B: The Memorandum: Technical Part 91
Appendix C: Derivation: The Critical Mass 97
Appendix D: Derivation: Weapon Yield 107
Appendix E: Equivalence of Peierls and Diffusion Theory Criticality Analyses in The Case of Large Critical Radius 117
Appendix F: Glossary of Symbols 125
Appendix G: Physical Constants and Conversion Factors 127

Chapter 1
Nuclear Fission: A Review

Abstract This chapter offers a review of the physics that underlies nuclear weapons, including a description of the discovery and verification of fission, the difference between reactors and bombs, why enriched uranium must be used for a bomb, and the concept of critical mass. All of these considerations had been established when Frisch and Peierls prepared their memorandum.

The Frisch-Peierls memorandum did not pop out of thin air; behind it lay years of research and discoveries, many serendipitous, that laid the basis for the discovery of nuclear fission and the possibility of nuclear weapons. This chapter gives a birds-eye view of this history from early 1934 onwards. The intent is not a detailed re-telling, but rather an overview of the most important developments. Material in this chapter is drawn from my books *The History and Science of the Manhattan Project* (2019), *Manhattan Project: The Story of the Century* (2020a), and *The Physics of the Manhattan Project* (2021a). A superb semi-popular treatment of the Manhattan Project that covers much of this background is Rhodes (1986); for a compact survey I humbly recommend Reed (2014). A superb technical history of Los Alamos can be found in Hoddeson et al. (1993).

1.1 Neutrons, Induced Radioactivity, and Transuranic Elements

We can begin this part of the story of the FP memorandum in Europe in early 1934. By that time, the structure of atoms as being akin to miniature solar systems with nuclear "suns" comprising protons and neutrons accompanied by orbiting electrons had been established: J. J. Thomson is credited with discovering the electron in 1897, Ernest Rutherford the proton and the concept of a nuclear atom about 1912, and his student, James Chadwick, with identifying the neutron as an independent element of nuclear structure in 1932. A brief reminder regarding the latter: Chadwick discovered the neutron as a consequence of what are now called (alpha, n) or (α, n) reactions, where an alpha-particle (identical to a nucleus of helium, ^4_2He) emitted in the natural decay of a radioactive element strikes a target foil of a light element such as beryllium

© The Author(s), under exclusive license to Springer Nature Switzerland AG 2025
B. C. Reed, *The Frisch-Peierls Memorandum*,
SpringerBriefs in History of Science and Technology,
https://doi.org/10.1007/978-3-031-95929-5_1

or aluminum, which causes a neutron to be released. In such experiments, the target foil has to be a light element; naturally-emitted alphas are not energetic enough to overcome the Coulomb-force repulsion of heavy-element target nuclei: You will not get an (α, n) reaction with a target element such as gold, silver, or lead. We will encounter (α, n) reactions again on various occasions.

In early 1934, Irène and Frédéric Joliot-Curie, working in Paris, were performing some follow-up experiments involving bombarding foils of aluminum with alpha particles emitted in the natural decay of a sample of polonium. To their surprise, their Geiger counter, which was being used to monitor the polonium source, continued to register a signal even after the source was removed. The signal decayed with a half-life of about 3 min. Performing the experiment in a magnetic field led them to conclude that positively-charged particles were being emitted *from the foil*, specifically, what are now known as positrons: Positive electrons. These are also known as β^+ particles. They proposed a two-stage reaction to explain their observations. First, that the release of a neutron in the (α, n) reaction was accompanied by formation of phosphorous-30, a conclusion dictated by conservation of electric charge and nucleon number:

$$^4_2\text{He} + ^{27}_{13}\text{Al} \rightarrow ^1_0\text{n} + ^{30}_{15}\text{P}. \qquad (1.1)$$

With the neutron emitted here, the resulting phosphorous-30 nucleus is a little proton-rich for its remaining number of neutrons, or, equivalently, a little neutron-poor for its number of protons. Nature deals with this by having one of the protons spontaneously transmute into a neutron. This however would imply the spontaneous loss of a positive charge; nature balances the books by simultaneously creating a positron, the positively-charged anti-particle of an electron, which is ejected to the outside world where it can be detected. A positron is also known as a β^+ particle; the "beta" terminology is a holdover from the early days of nuclear physics, when decays that were generating either positively or negatively-charged particles were being detected, but it was not yet realized that they were positrons or electrons. The result of this is that the phosphorous-30 nucleus transmutes to silicon; the modern value for the half-life is 2.5 min. The emitted positron is omitted here; it is the resulting nucleus that is important:

$$^{30}_{15}\text{P} \xrightarrow[2.5\,\text{min}]{\beta^+} {^{30}_{14}\text{Si}}. \qquad (1.2)$$

To check their interpretation, the Joliot-Curies dissolved the bombarded aluminum in acid; the small amount of phosphorous created could be separated and chemically identified as such. That the radioactivity "carried with" the phosphorous and did not remain with the aluminum verified their suspicion. Bombardment of other light elements showed similar effects. They reported their discovery in the January 15 edition of the journal of the French Academy of Sciences; an English version appeared in the February 10 edition of the British journal *Nature*. This discovery of artificially-induced radioactivity opened up the field of synthesizing short-lived isotopes for medical treatments from which millions of people have benefited.

1.1 Neutrons, Induced Radioactivity, and Transuranic Elements

Surprisingly, neither Chadwick nor the Joliot-Curies experimented with using the neutrons liberated in (α, n) reactions as bombarding particles themselves, likely on account of the fact that not many neutrons are actually produced. This notion did, however, occur to a young physicist at the University of Rome, Enrico Fermi; Fig. 1.1. Fermi had established himself as a first-rate theoretical physicist at a young age, publishing his first paper while still a student. In his early twenties he had prepared a review article on relativity theory; he also made seminal contributions to statistical mechanics and the theory of beta decay. He was to prove equally gifted as a nuclear experimentalist.

Fermi desired to break into nuclear experimentation, and saw his opening in the unexploited possibility of neutron bombardment. In the spring of 1934 he began work with a group of collaborators and students; by good fortune, his laboratory was located in the same building as the Physical Laboratory of the Institute of Public Health, which was charged with controlling radioactive substances in Italy. The Laboratory held many radium sources that had been used for cancer treatments. Radium decays naturally to radon gas, which, when mixed with powdered beryllium, gives rise to a copious supply of neutrons via (α, n) reactions as the radon itself undergoes alpha-decay. The radon and beryllium were placed in small glass vials; the resulting neutrons would emerge with energies of up to about 10 million electron-volts (MeV), more than energetic enough to escape through the thin walls of the vials, which were surrounded by a sample of the element to be bombarded. Fermi estimated that his sources yielded about 100,000 neutrons per second.

Fermi's goal was to see if he could induce artificial radioactivity with neutron bombardment. Possibly anxious to see if he could induce heavy-element radioactivity, his first target was platinum (atomic number 78), but no discernible signal was detected. Perhaps inspired by the Joliot-Curies' experience, he then turned to aluminum. Here he did succeed, and found a different half-life than they had. The reaction involved ejection of a proton (hydrogen nucleus) from the bombarded aluminum, leaving behind magnesium,

$$ {}_0^1 n + {}_{13}^{27}\text{Al} \rightarrow {}_1^1\text{H} + {}_{12}^{27}\text{Mg}. \tag{1.3} $$

Fig. 1.1 Laura and Enrico Fermi, 1954. Enrico is holding his trusty slide rule. *Source* Public domain; https://commons.wikimedia.org/wiki/File:HD.1A.020_(12750063023).jpg

With a proton ejected, the magnesium nucleus is a little neutron-rich for its number of protons. In analogy to the Joliot-Curie β^+ decay, the magnesium nucleus undergoes a form of beta-decay where one of its neutrons transmutes to a proton. But this would imply the spontaneous creation of a positive charge, so nature conserves charge by simultaneously creating an ordinary electron, which flies off to the outside world: A β^- particle. This process has a half-life of about 10 min:

$$^{27}_{12}\text{Mg} \xrightarrow{9.5\,\text{min}\;\beta^-} {}^{27}_{13}\text{Al}. \tag{1.4}$$

This first success occurred on March 20, 1934, and Fermi announced his discovery five days later in the official journal of the Italian National Research Council; an English-language report appeared in the May 19 edition of *Nature*. By late April, the Rome group had performed experiments on about 30 elements, 22 of which yielded induced beta-decay.

On moving up the periodic table to heavier-element targets, processes like this proved to be common, with neutron-rich products undergoing β^- decay. Bombardment of gold is a typical example (no proton gets ejected in this case):

$$^{1}_{0}\text{n} + {}^{197}_{79}\text{Au} \rightarrow {}^{198}_{79}\text{Au} \xrightarrow{2.69\,\text{days}\;\beta^-} {}^{198}_{80}\text{Hg}. \tag{1.5}$$

In this reaction, the first step is that the Gold-197 nucleus absorbs the incoming neutron to become Gold-198, which then decays due to being neutron-rich. By the early summer of 1934, Fermi had prepared improved sources, which he estimated were yielding about a million neutrons per second. Based on work with these, he published a stunning result in the June 16, 1934, edition of *Nature*: That his group was producing *transuranic* elements, that is, ones with atomic numbers greater than that of uranium. Since uranium was the heaviest-known element, this meant that they believed that they were synthesizing new elements. If true, this would be a remarkable development.

Fermi's assertion was based on the fact that uranium could be stimulated to produce beta-decay upon neutron bombardment. The results were somewhat confusing, however, with evidence for several half-lives appearing. Whether this was some sort of chain of decays or parallel sequences of decays was unclear, but, whatever was occurring, the initial step was presumably the formation of a heavy isotope of uranium followed by a beta-decay as in the gold reaction above:

$$^{1}_{0}\text{n} + {}^{238}_{92}\text{U} \rightarrow {}^{239}_{92}\text{U} \xrightarrow{\beta^-} {}^{239}_{93}\text{X}, \tag{1.6}$$

where X denotes a new, transuranic element. The half-life for the decay from ${}^{239}_{92}\text{U}$ to ${}^{239}_{93}\text{X}$ is about 23.5 min.

It proved possible to chemically separate the decay product X from the bombarded uranium, and analysis showed that it did not appear to be any of the elements between lead (atomic number 82) and uranium. Since no natural or artificial transmutation had ever been observed to change the identity of a target element by more than one or

1.1 Neutrons, Induced Radioactivity, and Transuranic Elements

two places in the periodic table, it would have seemed perfectly plausible to assume that a new element was being created. The product $^{239}_{93}$X itself undergoes β^- decay with a half-life of about 2.5 days.

Fermi's next discovery was serendipitous, but would prove key to the eventual development of plutonium-based nuclear weapons. In October 1934, he was conducting some calibration experiments which involved "filtering" neutrons, that is, reducing their energy by interposing layers of material, usually lead, between the neutron source and the target sample. But on one occasion he used a piece of paraffin (wax) instead, finding, to his surprise, that the presence of the paraffin caused the level of induced radioactivity to increase, sometimes dramatically. Further experimentation showed that the effect was characteristic of filtering materials that contained hydrogen; paraffin and water were most effective.

Within a few hours, Fermi developed a working hypothesis: That by being slowed by collisions with hydrogen nuclei, the neutrons would have more time in the vicinity of target nuclei to induce a reaction. Neutrons and protons have essentially identical masses, and, as with a billiard-ball collision, a neutron striking an initially stationary proton head-on would essentially be brought to a stop. Since atoms always have random motions due to being at a temperature that is above absolute zero, the incoming neutrons will never be brought to dead stops, but in practice only a few centimeters of paraffin or water are needed to bring them to an average speed characteristic of the temperature of the slowing medium. This process is now called "thermalization." Nuclear physicists define "thermal" neutrons as having kinetic energy equivalent to a temperature of 298 K, or 77 °F–just somewhat warmer than room temperature. The speed of a thermal neutron is about 2200 m per second, and the corresponding kinetic energy is about 0.025 eV, much less than the ~10 MeV of Fermi's radon-beryllium neutrons. Thermal neutrons are also known as "slow" neutrons; those of MeV-scale kinetic energies are, for obvious reasons, termed "fast." The water or paraffin is now known as a "moderator"; graphite, the material used in pencils, also works well in this respect.

The distinction between "fast" and "slow" neutrons is vital for understanding the Frisch-Peierls memorandum. When uranium is bombarded by neutrons, what happens depends very critically on the kinetic energies of the neutrons. Fast and slow neutrons lie at the heart of why nuclear reactors and bombs function differently, and why a bomb requires "enriched" uranium and fast neutrons to function. There will be much more said about fast and slow neutrons throughout the rest of this book.

Fermi's hypothesis that slower neutrons have a greater chance of inducing a reaction is now quantified in the concept of a reaction cross-section. This is a measure of the cross-sectional area that a target nucleus effectively presents to a bombarding particle that results in a given reaction. As you may know, cross-sections are designated with the Greek letter sigma (σ), equivalent to the English letter "s", which serves as a reminder that they have units of surface area. The fundamental unit of cross-section is whimsically termed a "barn," abbreviated as b or bn; 1 bn = 10^{-28} m^2. This miniscule number is characteristic of the geometric cross-sectional area of nuclei. Because of a quantum-mechanical effect known as the de Broglie wavelength, a slower bombarding particle will *appear* larger to a target nucleus than does a faster one, sometimes by factors of hundreds. Also, sometimes for a given target nucleus, different reactions can be induced; the possibilities are known as reaction "channels," and each channel will have its own characteristic run of cross-section as a function of bombarding-particle energy. We will have much more to say about cross-sections throughout the remainder of this book, and some plots of cross-sections for uranium under neutron bombardment appear later in this chapter.

Fermi was awarded the 1938 Nobel Prize for Physics for his demonstration of the existence of new radioactive elements produced by neutron irradiation. His wife and children were Jewish, and the family used the excuse of the trip to Stockholm to escape the fascist political situation in Italy by emigrating to America after the award ceremony; he had arranged for a position at Columbia University in New York.

Before proceeding to the story of the discovery of nuclear fission, an important intervening discovery needs to be mentioned. This is that uranium possesses a second, much less abundant isotope than the U-238 that Fermi had assumed was the sole form of that element. ^{238}U had been discovered by British physicist Francis Aston in 1931 as he was developing mass spectroscopy. In the summer of 1935, Arthur Dempster of the University of Chicago discovered evidence for a lighter isotope of mass number 235. Dempster estimated ^{235}U to be present to an extent of less than one percent of the abundance of its sister isotope of mass 238. The modern figures are that natural uranium ore comprises 99.3% ^{238}U and 0.7% ^{235}U. Within a few years, this fraction of a percent would change the world.

1.2 Discovery and Verification of Fission

Fermi's claim that transuranic elements could be created through neutron bombardment stimulated great interest within the nuclear research community. In addition to the Joliot-Curies, the other main leaders of that community were Otto Hahn and Lise Meitner at the Kaiser Wilhelm Institute for Chemistry in Berlin; Fig. 1.2. Hahn, a chemist specializing in the properties of radioactive elements, and Meitner, a physicist, had known each other and collaborated on-and-off for 30 years. In 1918, they discovered the rare element protactinium, and by the 1930s had accumulated years of experience with radioactive elements. Meitner became interested in Fermi's experiments, and convinced Hahn that they should investigate exactly how uranium behaved under slow neutron bombardment. To help with the necessary delicate chemical separation work, they brought on board chemist Fritz Strassmann.

To understand the Berlin group's assignments of identities for putative new elements, it is helpful to briefly digress into some chemistry. In the periodic table, elements in a given column behave similarly in their chemical properties. It was presumed that elements 93, 94, and so on would have chemical properties similar to the elements above them in the columns of the table in which they were expected to reside. Those elements are successively rhenium (above element 93), osmium (94), iridium (95), platinum (96), and gold (97). The anticipated new elements were given the tentative names eka-rhenium (EkaRe), eka-osmium, and so forth; "eka" is from

1.2 Discovery and Verification of Fission

Fig. 1.2 Otto Hahn and Lise Meitner in their laboratory, likely 1920's. *Source* Public domain; https://commons.wikimedia.org/wiki/File:Otto_Hahn_und_Lise_Meitner.jpg

the Greek for "beyond." In line with this expectation, Hahn, Meitner, and Strassmann separated their induced radioactivities from uranium by precipitating them out of solutions (the bombarded uranium had to be dissolved) with compounds containing elements in the presumed columns; these acted as "carriers" for the products being sought. This was an established technique they had perfected over many years. The work was demanding, however: The slight amounts of products created had to be thoroughly isolated from the bombarded uranium lest the natural radioactivity of the latter overwhelm the induced activity being sought.

By 1937 the situation had become very confusing: The Berlin team had identified no less than nine distinct half lives arising from uranium bombardment, many more than Fermi had detected. These were thought to arise from either radioactive isotopes of elements close to uranium (such as radium) or from new transuranic elements as far out as eka-Gold. Chemically, these identifications seemed secure, but Meitner struggled to understand the physics involved. How could the neutrons be initiating so many different processes in uranium, and how could such lengthy decay sequences arise? No other bombarded elements behaved in this way.

Tragically, Meitner's life was upended in mid-1938 when she was forced to flee Berlin. Born into a Jewish family in Austria, her Austrian citizenship had protected her from German anti-Semitic laws. That protection ended with the German annexation of Austria in March of that year. On July 13 she fled to Holland and then made her way to Sweden, where she was given a position at the Nobel Institute for Experimental Physics. She continued to collaborate with Hahn and Strassmann by letter, but her career was essentially destroyed.

In December 1938, Hahn and Strassmann came to a startling conclusion: Refinement of their chemical techniques indicated that neutron bombardment of uranium was revealing evidence for yielding barium as a reaction product. This was a dramatic surprise; the atomic weight of barium is only about half that of uranium. Could a uranium nucleus really be splitting into two halves?

On December 19, Hahn wrote to Meitner to seek her opinion as to what could be happening. The letter reached her in Stockholm on December 21. She replied at once: "Your ... results are very startling ... At the moment the assumption of such a thoroughgoing breakup seems very difficult to me, but in nuclear physics we have experienced so many surprises that one cannot unconditionally say: it is impossible." Hahn would receive her reply on December 23; he and Strassmann wasted no time in submitting a paper to the journal *Naturwissenschaften* reporting the discovery.

Also on December 23, Meitner traveled from Stockholm to spend Christmas with friends near Göteborg on the west coast of Sweden. Her nephew, Otto Frisch (yes, our Frisch), a nuclear physicist and another refugee from Austria, was then working at Niels Bohr's Institute for Theoretical Physics in Copenhagen. He traveled to Sweden to spend Christmas with his aunt, arriving also around the 23rd.

At some time during the following few days Meitner and Frisch went for a walk in the snow. She drew him into a discussion of Hahn's letter, and, as Frisch later related in his memoir, they sat down on a tree trunk and began to calculate on scraps of paper. Working from a theoretical model of nuclei that had been developed some years previously by George Gamow and Niels Bohr, the "liquid-drop" model (about which more below), Meitner and Frisch knew that uranium nuclei with their many protons are near the limit of stability beyond which no additional number of neutrons can inhibit them from spontaneously breaking up. Uranium nuclei are somewhat like wobbly drops, liable to fragment in response to a modest provocation such as the impact of a particle from the outside. If a uranium nucleus broke in two, the resulting fragments would experience a powerful mutually repulsive Coulomb force and fly away from each other at high speeds; this is depicted in Fig. 1.3.

Meitner had atomic mass data committed to memory, and calculated that in such a breakup, the fragments would total to a mass less than that of a uranium nucleus by about one-fifth of the mass of a proton, equivalent to an energy of about 170 MeV. This energy would appear to the outside world in the form of the kinetic energy of the fragments. Thus was the process of fission conceived in a snowy Swedish forest, a dual-edged Christmas present to the world if there ever was one.

The 170 MeV liberated in the fission of uranium is vastly greater than that in any chemical reaction, which is typically on the order a few eV. *On an atom-for atom basis, a nuclear reaction libertes millions of times as much energy as does a chemical reaction.* One gram of uranium contains some 2.5×10^{21} atoms, so fission of one kilogram of atoms will liberate some 4.4×10^{26} MeV, or 7×10^{13} J. Explosion of a thousand metric tons (10^6 kg; a kiloton) of TNT liberates some 4.2×10^{12} J; one

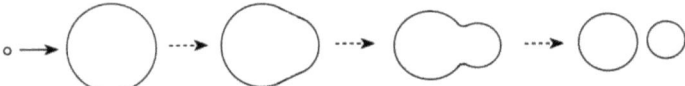

Fig. 1.3 Schematic representation of steps in the progression of fission. An initially spherical nucleus (left) is perturbed by a bombarding neutron, and begins to distort. Lobes form which force each other apart due to electrical repulsion, leading to fission. Sketch is not to scale. In reality, the volume of the nucleus is conserved in the process; the volumes of the final two spheres sum to that of the initial nucleus plus that of the incoming neutron. Sketch by author

1.2 Discovery and Verification of Fission

kilogram of uranium is thus potentially equivalent to about 17 kt of conventional explosive. It is no wonder that the prospect of such an immense release of energy caught the attention of physicists and eventually governments and military planners around the world. We will have more to say about kilotons later.

There are several further layers of this story yet to be revealed, but imagine for a moment Meitner's situation when faced with this stunning revelation. Most of the decay patterns which for years she had attributed to transuranic or near-to-uranium elements were actually the products of neutron-rich fission fragments decaying toward stability: Fallout.

In another of the curious confluences of events involved in nuclear history, it was around this time that Enrico Fermi and his family set out for America. He would know nothing of these developments until he met Niels Bohr three weeks later in New York.

On New Year's Day, Frisch returned to Copenhagen, keeping in telephone contact with his aunt as they drafted a paper based on the work they had begun a few days earlier. In his memoirs, Frisch relates that in all the excitement, he and Meitner missed an important point: The possibility of a chain reaction. A colleague suggested to him that the fission fragments might contain enough energy to each eject a neutron or two, which might go on to cause other fissions. That the fragments would be neutron-rich in comparison to stable nuclei of the same atomic number made this possibility very real. Frisch's immediate response was that if such were the case, no deposits of uranium ore should exist as they would have blown themselves up long ago. But he then realized that this argument was too simplistic: Ores contained many other elements which might capture neutrons, and many neutrons could escape before causing another fission. A chain reaction might not be a fantasy.

In early January 1939, the focus of fission research shifted briefly to Copenhagen, and then primarily to America. On the third, Frisch caught up with Niels Bohr to relate the situation. The conversation was brief; Bohr was preparing to spend a semester at the Institute for Advanced Study in Princeton, New Jersey. According to Frisch, Bohr's reaction was to wonder why he had not thought of fission before; Frisch would later depict Bohr as hitting himself on the forehead and exclaiming, "Oh what idiots we have all been. Oh but this is wonderful! This is just as it must be! Have you and Lise Meitner written a paper about it?" Bohr promised not to disclose the discovery until their paper had been prepared.

Frisch also discussed the situation with a theoretician colleague, George Placzek, who encouraged him to set up an experiment to detect the expected high-energy fission fragments. He did so on Friday, January 13, and immediately detected the fragments, becoming the first person to set up an experiment designed to deliberately demonstrate and detect fission. He is also credited with coining the term "nuclear fission," after having asked a biologist what term was used for the process of cell division: "binary fission." Frisch also tested the element thorium, which proved to act like uranium in that it would fission under bombardment by fast (unmoderated) neutrons, but to act unlike uranium in that it did not do so when bombarded with slow (moderated) neutrons. This asymmetry would prove to be a crucial observation a few weeks later as Bohr worked to understand the underlying physics of the process.

Frisch prepared two papers for submission to *Nature*; the first was co-authored with Meitner and described their Christmastime insights, while the second described his experiments. The joint paper was published on February 11, and the experimental one on February 18 [Meitner and Frisch (1939); Frisch (1939)]. Frisch published an account of his fateful walk in the snow with his aunt and a reminiscence of her in the last decade of his life; Frisch (1973, 1978).

Bohr sailed to America, accompanied by collaborator Léon Rosenfeld of the University of Liège in Belgium. Bohr had a blackboard installed in his stateroom, and during the voyage he and Rosenfeld began to develop a theoretical understanding of fission. They arrived in New York on January 16, where they were met by Enrico and Laura Fermi. Bohr remained in New York, but Rosenfeld left for Princeton. However, Bohr had not told Rosenfeld to keep the news quiet until Meitner and Frisch's paper had been submitted, and Rosenfeld let the cat out of the bag that evening at a meeting of the Princeton Physics Journal Club. When Bohr learned this, he hastily drafted his own note to Nature on January 20 to assert Meitner and Frisch's priority; it would be published on February 25 (Bohr 1939a).

Bohr's paper gave the first theoretical speculations on the details of fission, and is worth comment. In it, he gave a description of how the fission process could be envisioned based on his liquid drop nuclear model that had guided Meitner and Frisch's thinking. Bohr speculated that in ordinary non-fission reactions, the energy of the bombarding particle was distributed in the target nucleus among various modes of vibration in a manner resembling the agitation of a liquid drop. If a large fraction of the energy should come to be concentrated on some particle at the surface of the nucleus–say a proton, neutron, or alpha particle—then that particle will be ejected. In a fission reaction however, Bohr reasoned that the distribution of energy would have to result in a mode of vibration of the nucleus that involved a deformation of the surface (Fig. 1.3), speculating that in a heavy nucleus the energy necessary to distort the surface to the point of leading to fission must be of the same order of magnitude as that needed to cause the escape of a single particle from a lighter nucleus. The concept of a requisite deformation energy would soon find more rigorous quantitative expression as the "fission barrier," which is described later in this chapter.

The first demonstration of fission in America occurred at Columbia University. On January 25, Bohr, while on his way to attend a conference in Washington, stopped at Columbia to see Fermi. Fermi was out, but Bohr met one of his graduate students, Herbert Anderson. Bohr related the news; Anderson, who was preparing a thesis on neutron scattering, instantly understood the significance of the discovery, and that evening set up an experiment to detect fission fragments with an ionization chamber he had prepared for his thesis work. The news was telegraphed to Fermi, who was already in Washington.

What drew Bohr and Fermi to Washington was the Fifth Washington Conference on Theoretical Physics, which was being held at George Washington University. These meetings were organized by George Gamow and Edward Teller (of later hydrogen bomb fame), both of whom were then at GWU. The topic of the meeting was to be low-temperature physics, but that agenda soon changed. Gamow opened by introducing Bohr, who related Hahn and Strassmann's discovery and Meitner and

1.2 Discovery and Verification of Fission

Frisch's interpretation. The news electrified the fifty-odd participants, some of whom left to perform their own experiments. Within a few days, the effect was verified at Johns Hopkins University in Baltimore, the Carnegie Institution in Washington, and the University of California at Berkeley, Robert Oppenheimer's home base. The *New York Times* reported the discovery in its Sunday, January 29, edition, noting that scientists at the meeting thought that it might be twenty or twenty-five years before the phenomenon could be put to use.

Fission can happen in a number of ways, but it is always accompanied by a tremendous release of energy and the prompt emission of two or three "secondary" neutrons; an example is the Hahn and Strassmann discovery reaction,

$$^{1}_{0}n + ^{235}_{92}U \rightarrow ^{141}_{56}Ba + ^{92}_{36}Kr + 3\left(^{1}_{0}n\right) + 170 \text{ MeV}. \tag{1.7}$$

A sharp-eyed reader will note that this reaction involves the 235 isotope of uranium, not the more common 238 isotope. This is an important point; much of what follows explains why this is.

The vast majority of the liberated energy is carried off in the form of kinetic energy by the barium and krypton fission products, but the neutrons carry off on average about 2 MeV each, a number that will prove to be important. The neutron-rich fission products then decay by a series of beta decays,

$$^{141}_{56}Ba \xrightarrow{18.3 \text{ min}} ^{141}_{57}La \xrightarrow{3.9 \text{ hr}} ^{141}_{58}Ce \xrightarrow{32.5 \text{ days}} ^{141}_{59}Pr \tag{1.8}$$

and

$$^{91}_{36}Kr \xrightarrow{8.6 \text{ sec}} ^{91}_{37}Rb \xrightarrow{58 \text{ sec}} ^{91}_{38}Sr \xrightarrow{9.5 \text{ hr}} ^{91}_{39}Y \xrightarrow{58.5 \text{ days}} ^{91}_{40}Zr. \tag{1.9}$$

As some 30 different elements are produced by uranium fission, it is no wonder that Hahn, Meitner, and Strassmann had observed a confusing mixture of decay chains.

Obviously, one has to have on average at least one neutron liberated per fission if there is to be any hope of achieving a neutron-moderated chain reaction. Soon after the discovery of fission, a number of research teams began looking for evidence of secondary neutrons, and proof of their presence was not long in coming. On March 16, two independent groups at Columbia submitted reports to *Physical Review* announcing their discovery: Anderson et al. (1939b), and Szilard and Zinn (1939). Both groups estimated about two neutrons emitted per each captured; their papers were published on April 15. Szilard and Zinn configured their experiment to detect the emission of *fast* neutrons as a consequence of fission induced by slow neutrons, and indeed observed them. Szilard later recalled his reaction upon detecting the neutrons: "That night, there was very little doubt in my mind that the world was headed for grief." The modern value for the average number of secondary neutrons liberated by ^{235}U when fissioned by thermal neutrons is about 2.4. We will have more to say regarding this number when considering Frisch and Peierls' work.

The physics of fission is a complex topic, and is the subject of the following two sections.

1.3 Nuclear Parity and Neutrons Fast and Slow

Otto Frisch's observation that the likelihood of uranium to fission depended on the velocity of bombarding neutrons and that uranium and thorium differed in their responses to slow-neutron bombardment catalyzed several crucial revelations on the part of Niels Bohr in early 1939.

Sometime in January, George Placzek arrived at Princeton. Over breakfast with Bohr and Rosenfeld one morning, the conversation turned to fission. When Bohr expressed relief that physics was now rid of purported transuranic elements, Placzek protested, arguing that the situation was more confused than ever. He pointed out that both uranium and thorium were known to have strong propensities to capture slow neutrons and subsequently decay. Did this mean that transmutations and energy-releasing fissions were somehow occurring simultaneously? Also, why did uranium fission under slow-neutron bombardment but thorium not? Bohr then had his epochal revelation.

Working with remarkable haste, he prepared and sent off a paper to the *Physical Review*, which was published in the February 15 edition (Bohr 1939b). In this paper, he developed arguments to show that it was likely the rare isotope ^{235}U that must be responsible for slow-neutron fission in that element, and to explain why thorium did not exhibit slow-neutron fission.

Bohr's argument comprised two components. The first involved the liquid-drop model discussed above. Bohr linked this argument to earlier experiments of Meitner, Hahn, and Strassmann wherein they examined the neutron-capture response of uranium to neutrons of varying speeds; this was often referred to as radiative capture, as the struck nucleus sheds excess energy in the form of a gamma-ray after the collision. When plotted as a graph of capture cross-section versus neutron energy, this work had revealed a rich forest of what were called "resonance capture lines" for neutrons of energies of from a few to thousands of eV, that is, energies at which capture was especially probable; look ahead to Fig. 1.4. Based on arguments from statistical mechanics, Meitner, Hahn, and Strassmann had concluded that these resonances were likely attributable to the abundant isotope, ^{238}U. However, these resonance captures were not associated with any corresponding increase in the fission probability, which led Bohr to infer that ^{238}U nuclei must be very stable against even medium-energy neutron bombardment. If ^{238}U does not fission under medium-energy neutron bombardment, one would certainly not expect it to do so under slow-neutron bombardment. (Spoiler alert: This statement will be revisited later.) Thus, Bohr reasoned that this left ^{235}U as the only candidate for slow-neutron fission.

Bohr's second argument helped to clarify what was happening with thorium by looking at the situation from the point of view of what is known as "nuclear parity," a term used to designate the evenness or oddness of the number of protons and

1.3 Nuclear Parity and Neutrons Fast and Slow

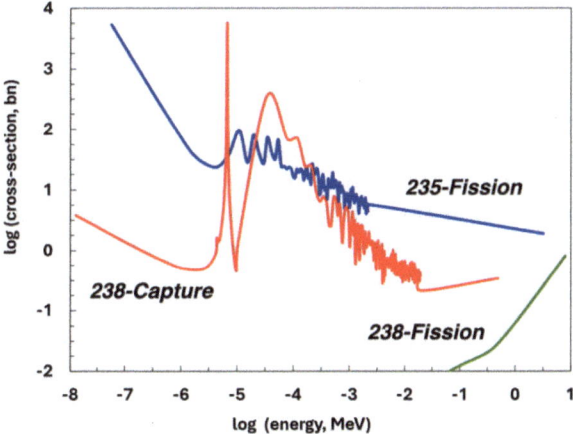

Fig. 1.4 Approximate cross-sections (barns) for ^{235}U and ^{238}U vs. bombarding neutron energy in MeV. The data have been heavily averaged across the resonance peaks to simplify the plot. Scales are logarithmic (base 10). This plot is intended to be used only for comparative purposes and should not be used to infer the precise cross-section at any particular energy. For example, for thermal neutrons with $\log_{10}$(energy) = –7.6, the fission cross-section for ^{235}U is 585 bn. The fission cross-section for ^{235}U increases rapidly for lower energies, which skews the average around this energy high in this figure. Plot prepared by author

Table 1.1 Neutron number information

Isotope	Number of neutrons	Energy liberated on neutron capture, MeV (approx.)
$^{238}_{92}$U	146	5
$^{235}_{92}$U	143	6.5
$^{232}_{90}$Th	142	5

neutrons that nuclei contain. For many years before the discovery of fission, empirical evidence indicated that nuclei which contain even numbers of protons and neutrons ("even parity") are more stable than those with odd numbers of either or both. Since fission is a neutron-induced phenomenon, our concern here is with the evenness or oddness of the numbers of neutrons involved.

Table 1.1 summarizes the situation for uranium and thorium isotopes. $^{235}_{92}$U has an odd number of neutrons (143), while $^{238}_{92}$U's neutron complement is even (146), as is that for $^{232}_{90}$Th. We will get to the meaning of the last column of the table presently.

Bohr pointed out that uranium consists of two isotopes, one with an even number of neutrons and one with an odd number, whereas thorium has only one stable isotope, which has an even number. If it is the odd-parity isotope of uranium that is responsible for slow-neutron fission, then one might expect *not* to see slow-neutron fission in thorium as it lacks an isotope of such neutron parity. This was consistent with what Otto Frisch and others had observed.

The second-last paragraph of Bohr's paper presented a hypothesis concerning fast-neutron fission, a speculation which seems to have been largely overlooked at the time with all the attention being devoted to slow neutrons. It is worth examining this part of his argument in some detail.

Quantum-mechanical considerations indicated that as the energy of bombarding neutrons increases (that is, as they become faster), the fission cross-section should generally decrease because of the decreasing de Broglie wavelength. For very fast (MeV) neutrons, the cross-section should never exceed the geometric cross-section of the nucleus itself, which for uranium is about 1.7 barns; this is an important point that will reappear in the discussion of the Frisch-Peierls memorandum in section 4.5.2. Since ^{238}U did not fission under intermediate-energy neutron bombardment, it would certainly not be expected to do so when struck by fast ones because of the lower cross-section to be expected at higher energies. Now, this reasoning appears to contradict the claim above that ^{238}U should not suffer slow-neutron fission, but be patient: There is more to come. On the other hand (as Bohr pointed out), ^{235}U might have a chance of sustaining fast-neutron fission in view of its apparently very large cross-section for slow neutrons. That is, might there be sufficient "remaining" cross-section for fast neutrons despite the expected decrease in cross-section with increasing neutron energy? This was speculation as yet, but by no means out of the question.

The levels of argument that Bohr developed in a two-page paper are impressive. In his own words, it amounted to (this author's additions in parentheses) "allowing us to account both for the observed yield of the process concerned for thermal neutrons and for the absence of any appreciable effect for neutrons of somewhat higher velocities. For fast neutrons ... because of the scarcity of the isotope concerned (that is, ^{235}U) the fission yields will be much smaller than those obtained from neutron impacts on the abundant isotope (^{238}U)." The details of his analysis would be revised as further experimental data accumulated, but by the spring of 1939 the general outlines of understanding of the response of different uranium isotopes to neutron bombardment and the prospects for a chain reaction were beginning to clarify.

What are the numbers listed in the last column of Table 1.1? These are of empirical origin, but are described here with a fanciful analogy.

Imagine a nucleus as a party; the guests are individual nucleons. Again, our concern is with the neutrons. A neutron from outside tries to crash the party. Nucleons already present can be thought of as each willing to give up a small amount of their mass to make room for an incoming neutron. Particularly preferable would be an incoming neutron whose addition would make the total number of neutrons even, the empirically more stable situation. For heavy nuclei, measured nuclear masses indicate that the already-present guests are collectively willing to sacrifice an amount of mass equivalent to about 6.5 MeV of energy to achieve an even number of neutrons. That liberated energy appears in the form of excitation energy of the nucleus; the party becomes louder, and some of the nucleons might fission out the door. On the other hand, a newcomer whose addition would make the total number of neutrons odd is also welcome (neutrons never repel other nucleons), but less so in that the nucleons already present are a little less willing to make room for an odd-one-out.

1.3 Nuclear Parity and Neutrons Fast and Slow

In this case they are willing to sacrifice only about 5 MeV mass equivalent, and the nucleus becomes less roiled than if it had sustained an odd-to-even neutron-number transition.

This scheme predicts that a nucleus which transforms from an odd to an even number of neutrons by capturing a neutron will liberate more energy than one that goes from an even to an odd number of neutrons, with the difference being about 1.5 MeV. This is what happens when a ^{235}U nucleus (initially an odd number of neutrons) takes in a neutron versus what happens when a ^{238}U nucleus (initially even) does so. It turns out that the extra 1.5 MeV is enough to cause a bombarded ^{235}U nucleus to fission, whereas a ^{238}U nucleus simply captures the incoming neutron, becomes agitated, and subsequently beta-decays. These energies liberated upon neutron capture are listed in the last column of Table 1.1. The exact numbers for ^{238}U, ^{235}U, and ^{232}Th are respectively 4.81, 6.55, and 4.79 MeV.

In short, when an odd-neutron nucleus such as ^{235}U takes in a neutron, it will find itself in a more excited energy state–and hence more prone to fission–than would an even-neutron one. Further numerical details will follow shortly and in the next section.

Even for nuclear physicists, these parity arguments are still largely in the realm of empirical numbers. At present, particle physics can only just predict the masses of individual fundamental particles from theories of the underlying physics of nuclear forces, let alone the mass of an entire heavy nucleus.

To close this section, some numbers. Look at Figs. 1.4 and 1.5. Figure 1.4 shows the neutron capture cross section for ^{238}U (red curve), the fission cross section for ^{238}U (short green curve, lower right), and the fission cross section for ^{235}U (blue curve) as a function of bombarding neutron energy in MeV. Cross-sections are in barns; both scales are logarithmic to accommodate the tremendous ranges involved. So far as Frisch and Peierls are concerned, this is to some extent cheating: This graph is based on present-day data. Modern tabulations of cross sections can run to tens to hundreds of thousands of lines, which are required to have the necessary resolution in energy to map the thousands of resonance peaks that occur. So much data would be overwhelming to graph in detail, so to prepare this plot I have averaged such files in groups of 100 lines of data; the resonance peaks get smeared together by the averaging but are still visible. Remember that in natural uranium, ^{238}U nuclei outnumber ^{235}U nuclei by a ratio of over 100-to-1; while the red and blue curves appear to be of roughly equal magnitude, you have to imagine multiplying the red one by a factor of 100 to account for this, that is, shift it up vertically by +2 in logarithm. This will emphasize that in ordinary uranium, capture of incoming neutrons by ^{238}U nuclei vastly overwhelms the probability of fission by ^{235}U nuclei. ^{238}U nuclei fission only under bombardment by very energetic neutrons, which corresponds to the short green curve in the lower right corner of the plot. Figure 1.5 shows a detailed view of the ~0.03–3 MeV energy range with the addition of the cross-section for inelastic scattering from ^{238}U; this will come up in the discussion of the technical part of Frisch and Peierls' memorandum.

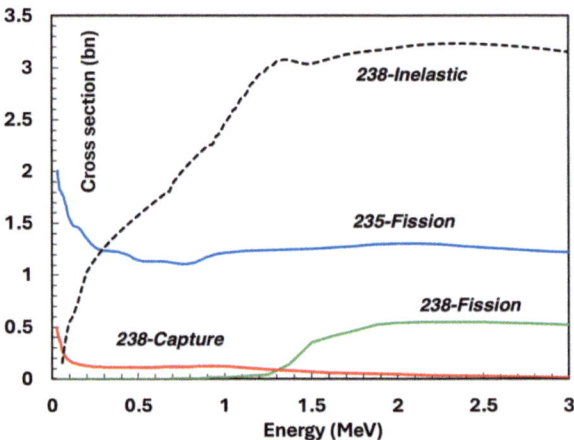

Fig. 1.5 High-energy cross-sections (barns) for neutron bombardment of ^{235}U and ^{238}U. Scales are linear. Color scheme as Fig. 1.4 with addition of the black dashed line for inelastic scattering of ^{238}U. The curves terminate at about 0.03 MeV on the left side to avoid the resonance spikes evident in Fig. 1.4

1.4 Bohr and Wheeler: The Fission Barrier and Chain Reactions

Niels Bohr's insight that it was likely the rare isotope ^{235}U that was responsible for slow-neutron fission was an important first step in an extensive chain of experimental and theoretical investigations into the fission process that unfolded over the following year. This work culminated with verification of his hypothesis at just about the time Frisch and Peierls were preparing their memorandum. These developments are surveyed in this section. For readers seeking more detail, a superb summary of developments to late 1939 can be found in a review paper by Turner (1939).

Upon arriving in America, Bohr began collaborating with John Wheeler, a young Assistant Professor at Princeton University. Bohr and Wheeler had known each other since 1934, when Wheeler spent a postdoctoral year with Bohr in Copenhagen. In the September 1, 1939, edition of the *Physical Review*, they published an extensive analysis of the energetics of fission (Bohr and Wheeler 1939). Their paper was lengthy and dense, but for the present purposes their conclusions can be summarized as follows. There are two main points.

Point 1. Bohr and Wheeler's analysis indicated that any otherwise stable nucleus can be induced to fission under neutron bombardment. However, any specific isotope possesses a characteristic *fission barrier* or, synonymously, *activation energy*, that is, a certain minimum amount of energy has to be supplied to deform the nucleus sufficiently to induce the process sketched in Fig. 1.3 to begin. This energy can be supplied by a combination of two contributions: (i) in the form of kinetic energy carried in by the bombarding neutron, and (ii) any binding energy liberated as described above when the target nucleus absorbs the bombarding neutron and becomes a different nuclide with its own characteristic mass. Both factors play roles in understanding uranium fission.

1.4 Bohr and Wheeler: The Fission Barrier and Chain Reactions

Bohr & Wheeler's analysis showed that fission barriers are immense for elements across most of the periodic table, in the tens of MeV, but decline to only a few MeV for the heaviest elements, which are somewhat unstable due to their large numbers of protons. Mid-weight elements can be fissioned, but require bombardment by extremely energetic neutrons, much more energetic than those typically released in fissions. But for heavy elements, the energy liberated upon neutron capture can rival the fission barrier.

Point 2. In 1936, Bohr had developed a conceptual model of nuclear reactions that is now known as the "compound nucleus" model. The motivation for this emerged from an experimental dilemma: When nuclei were bombarded with neutrons, the latter, even if slowed, should be moving fast enough to sail right through a nucleus with little chance of being captured. But captured they evidently were, with the nucleus then ejecting a neutron, proton, alpha-particle, or gamma-ray—but only after a time much longer than it would have taken the bombarding neutron to pass through. Bohr realized that the bombarding neutron would collide with one of the protons or neutrons within the target nucleus, and that those nucleons would then collide with others, dissipating the energy of the neutron. The struck nucleus would become agitated—in a sense, "hot"—and then, after some time, excess energy would be shed in the form of a gamma-ray or perhaps a particle near the surface of the nucleus would get ejected if it happened to acquire enough energy to escape. Based on this picture, Bohr and Wheeler speculated that fission is not an instantaneous process, but rather that the incoming neutron and target nucleus first combine to form an intermediate compound nucleus. Two cases are relevant for uranium:

$$^{1}_{0}n + ^{235}_{92}U \rightarrow ^{236}_{92}U \qquad (1.10)$$

and

$$^{1}_{0}n + ^{238}_{92}U \rightarrow ^{239}_{92}U. \qquad (1.11)$$

In accordance with the neutron parity changes described above, the energies liberated in these reactions are respectively 6.55 and 4.81 MeV. If the bombarding neutrons are "slow," that is, if they bring essentially no kinetic energy into the reactions, then the nucleus of ^{236}U formed in reaction (1.10) will find itself in an excited state with an internal energy of about 6.6 MeV, while the ^{239}U nucleus formed in reaction (1.11) will have a like energy of about 4.8 MeV. In comparison, the fission barriers for bombardment of ^{235}U and ^{238}U are respectively about 5.0 and 6.2 MeV; see Möller et al. (2015). *It is the differences between the binding energies liberated and the necessary activation energies that are crucial here.* In the case of ^{235}U, the liberated energy exceeds the activation barrier by nearly 1.5 MeV: *Any bombarding neutron, no matter how little kinetic energy it has, can induce fission in* ^{235}U. On the other hand, the energy liberated in reaction (1.11) falls some 1.4 MeV short of the fission barrier, which means that to fission ^{238}U by neutron bombardment requires supplying neutrons of at least this amount of energy. ^{235}U is known as a "fissile" nuclide, while ^{238}U is termed "fissionable".

The issue of the unsuitability of ^{238}U as a weapons material is actually somewhat more subtle than the above argument lets on, and is worthy of a digression. The average kinetic energy of secondary neutrons liberated in the fission of uranium nuclei is about 2 MeV, and about half of them have energies greater than the $\sim$1.4 MeV excitation energy of the n + ^{238}U $\rightarrow$ ^{239}U reaction. In view of this, it would appear that ^{238}U might make a viable weapons material. Why does it not? The problem depends on what happens when fast neutrons strike ^{238}U nuclei. This discussion assumes some modern numbers that Frisch and Peierls did not have access to, but even the information available at the time led them to discard consideration of ^{238}U.

When a neutron strikes and is scattered by a target nucleus, that is, if the neutron is "deflected" and goes on its way as opposed to being absorbed or causing a fission, the collision can happen in one of two ways: elastically or inelastically. In elastic scattering, the kinetic energy of the incoming neutron is essentially unaffected. But if the collision is inelastic, the neutron loses considerable kinetic energy; the "lost" energy goes into leaving the struck nucleus in an excited energy state, analogous to a chemical reaction that leaves an electron in a higher-energy orbit.

The problem with ^{238}U is twofold. Modern cross-section data reveal that a fast neutron striking a ^{238}U nucleus is about eight times as likely to be inelastically scattered as it is to induce a fission. However, as a result of such an interaction, the kinetic energy of the neutron is reduced to below the 1.4-MeV fission threshold. At lower energies, ^{238}U nuclei readily capture neutrons, taking them out of circulation for maintaining any chain reaction; see the red curve in Fig. 1.4. It is this inelastic scattering effect that resolves the apparent contradiction involving neutron speeds in the preceding section.

Despite its non-fissility, ^{238}U did play a crucial role in the Manhattan Project in breeding plutonium in nuclear reactors. This is an extensive story in its own right, but will be forgone here as Frisch and Peierls did not consider plutonium in their analysis.

To summarize to this point: First, consider trying to establish a chain-reaction using slow neutrons. The neutrons emitted in fissions will be fast, but are subject to the ^{238}U inelastic-scattering and capture problem described above. To have any hope of keeping the reaction going, the fission-liberated neutrons have to be slowed in order to (i) avoid being captured by ^{238}U nuclei while (ii) taking advantage of the enormous fission cross-section of ^{235}U for thermal neutrons. However, a bomb based on such a scheme would weigh tons and be impractical to deliver to a target in any way; essentially, it would be a reactor. Also, the neutrons would be so slow that the reaction would grow at a rate not much faster than an ordinary chemical reaction. The result would be that the device would heat itself, melt, and disperse, which would allow neutrons to escape and cause the reaction to shut down. A slow-neutron bomb would create an expensive fizzle. To create a reaction violent enough to warrant making a bomb requires using fast neutrons. In this case, the only isotope that might be able to sustain a chain reaction is ^{235}U, but this would require separating the two uranium isotopes in kilogram quantities. Even if the separation could be achieved, there was no guarantee in 1939 that some unanticipated effect might not arise that could render a bomb unworkable. It is not at all surprising that Niels Bohr thought that a weapon based on uranium fission would be impractical or impossible; this point will arise again later.

A few more comments on reactors versus bombs are appropriate here. As described above, ^{235}U nuclei have an enormous probability for undergoing fission when struck by slow neutrons: over 200 times the slow-neutron capture probability of ^{238}U. This factor is large enough to compensate for the small natural abundance of ^{235}U to the extent that in a sample of natural-abundance uranium, a slow neutron is about as likely to fission a nucleus of ^{235}U as it is to be captured by one of ^{238}U. This behavior is what makes possible the slow-neutron chain reactions used in nuclear reactors. *In effect, slowing fission-liberated neutrons is equivalent to enriching the abundance percentage of* ^{235}U. To repeat: If a neutron emitted in a fission can be slowed, then it has about as good a chance of going on to fission another ^{235}U nucleus as it does of being uselessly captured by a nucleus of ^{238}U. In actuality, both processes proceed simultaneously within a reactor: ^{235}U fissions generate energy and liberate neutrons, while ^{238}U nuclei capture some of the neutrons and become a waste product, which in fact decays to plutonium.

The trick to slowing neutrons during the very brief interval between when they are emitted in fissions and when they strike other nuclei is to work not with a single large mass of uranium but rather to disperse it as small chunks within a surrounding medium which slows neutrons without capturing them. The medium is known as a moderator, and the entire package is a reactor. During the Manhattan Project, the synonymous term "pile" was used in the literal sense of an arrangement of slugs of uranium metal embedded within a heap of moderating material. Ordinary water can serve as a moderator, but water does capture some neutrons and so requires using fuel enriched to a few percent ^{235}U. An excellent alternative moderator is heavy water, where the hydrogen atoms are already bonded with neutrons: $^{2}_{1}$H; because neutrons are already present in the moderating material, fission-liberated ones are less likely to be captured and taken out of circulation. Another excellent alternative is graphite, the material used in pencils; this was much more readily available at the time. By using moveable rods of a neutron-capturing material placed in the pile, the reaction can be controlled by inserting/withdrawing them as necessary. It is in this way that natural-abundance uranium proved capable of sustaining a controlled nuclear reaction, although not an explosive one. All commercial power-producing reactors operate via slow-neutron chain-reactions, although controlled fast-neutron designs are used in space-critical applications such as submarines. As described above, slow-neutron reactors cannot behave like bombs; the reaction is far too slow, and even if the control rods are rendered inoperative, the reactor will melt rather than blow up. The reactor explosion at Fukushima, Japan, was actually a steam explosion involving the reactor's cooling water, not a nuclear explosion.

1.5 Criticality

Following the discovery that uranium fission does give rise to secondary neutrons, researchers began to consider the conditions necessary for achieving a chain reaction, at least in theory.

Even if one has a fission to begin a putative chain reaction, the secondary neutrons that are liberated are by no means guaranteed to strike other nuclei; some will inevitably reach the surface of the sample and escape, particularly if it is small. As the size of the sample increases, the probability that a given neutron will escape decreases, and while the probability never goes strictly to zero for a finitely-large sample, it will eventually become low enough that a neutron is more likely to cause a subsequent fission than it is to escape. The key concept is that of a critical mass: The minimum mass of uranium that has to be assembled in one place in order to have a self-sustaining reaction which in principle continues until all of the uranium has fissioned, or, more likely, heats itself up and disperses. Here ^{235}U is had in mind; initially, some researchers were not clear on the roles of the isotopes. Technically, criticality is said to occur if the number of neutrons within the sample is increasing with time. Whether or not this can occur depends on the density of the fissile material, its cross-sections for fission and scattering, and the number of neutrons emitted per fission. To analyze the evolution of the number of neutrons in a reactor or bomb core requires the use of time-dependent diffusion theory, which is covered in a number of technically-oriented texts; for the present purposes the essential analysis is summarized in Appendix C. Diffusion theory goes back to classical thermodynamics, and was a well-established branch of theoretical physics by 1939.

The first attempt at computing a critical mass was published by French physicist Francis Perrin in the May 1, 1939, edition of the journal *Comptes Rendus*. Perrin applied diffusion theory to an assemblage of natural-abundance uranium, assuming fast (unmoderated) neutrons. With rough estimates for some of the relevant parameters, he arrived at a critical mass of 40,000 kg, an enormous amount. Perrin's result had no real relevance for a bomb, where pure ^{235}U is used. However, he did establish the relevant diffusion physics.

For the present purposes, the key early criticality paper is one published by Peierls in October 1939; Peierls (1939). In his memoirs, Peierls describes how he read Perrin's paper, and refined the calculation to consider a pure fissile isotope. Given the potential military applications, he had some doubts about openly publishing his analysis, and consulted Frisch on the advisability of doing so. Confident that Bohr had shown that an atomic bomb was not a realistic proposition, Frisch saw no reason not to publish. A few months later the two would find themselves in a very different position.

Peierls' analysis is mathematically complicated, but the essential result is that he developed explicit formulae for estimating the critical mass in two extreme cases. These were for when the number of neutrons generated per fission is close to one (large critical radius), and much greater than one (small). In the region of practical interest where the number is about 2.5, the two expressions turn out to not differ

1.6 Bohr Verified

Bohr's suggestion that ^{235}U that was responsible for slow-neutron fission–and the implications for fast-neutron fission—begged for experimental test. This meant isolating samples of ^{235}U and ^{238}U and subjecting them to neutron bombardment. The only practical method of isotope separation known at the time was mass spectroscopy, and the task of preparing the samples came into the hands of a master practitioner of that art, Alfred Nier of the University of Minnesota.

Nier had come to the attention of uranium physicists with a paper he had published in early 1939 in wherein he reported discovering a third, very rare, isotope of uranium, ^{234}U. This isotope is present to the extent of only about one atom per every 18,000 of ^{238}U in a sample of natural uranium, but Nier's mass spectrometer was sensitive enough to achieve the detection. ^{234}U plays no role in the story of the Frisch-Peierls memorandum, but Nier met Enrico Fermi at an American Physical Society meeting in Washington in April 1939, at which time Fermi encouraged him to try to separate small samples of the 235/238 isotopes. Busy with teaching, Nier did not take up the challenge until prodded again by Fermi in October of that year.

In order to achieve sufficient separation, Nier had to build a new mass spectrometer, which he completed in February 1940. His first successful separation runs were carried out on February 28 and 29. He glued the minute samples to a letter which he posted by airmail special delivery to Columbia University, where they were subjected to slow-neutron bombardment.

Nier's samples were truly miniscule. He did two separation runs, of 10 and 11 h durations, which he predicted yielded 0.17 and 0.29 *micro*grams of ^{238}U, respectively, assuming that all of the ions stuck to the collector. The amounts of ^{235}U would have been 1/140 as much, or about 1.2 and 2.1 *nano*grams. To collect a full kilogram at a rate of 2.1 nanograms per 11 h of operation would require some 600 million years of continuous operation, further testament to Niels Bohr's opinion of the impracticality of a fission bomb.

At Columbia, the ^{235}U samples clearly showed evidence for slow-neutron fission, while the ^{238}U samples showed none at all. Despite the minute samples, the Columbia team was able to estimate the thermal-neutron fission cross-section for ^{235}U as 400–500 barns; the modern value is 585. These results were reported in a paper published in the March 15, 1940, edition of the *Physical Review*; Nier et al. (1940). The paper closed with the observation that "These experiments emphasize the importance of uranium isotope separation on a larger scale for the investigation of chain reaction possibilities in uranium." Unfortunately, Nier's samples were too small to test for fast-neutron fission. A follow-up paper published a month later reported further results based on larger samples; this verified that ^{238}U fissioned fission only under fast-neutron bombardment. The paper did not report the energy of the fast neutrons; it must have been greater than the $\sim$1.4-MeV threshold discussed above. Slow-neutron fissility of ^{235}U was again verified, but the sample of 235 was too small to test for fast-neutron fission. However, the implication was clear: If ^{235}U had any appreciable fast-neutron cross-section, the possibility of an explosive chain reaction could be very real.

Niels Bohr's skepticism that the difficulty of separating isotopes would obviate the possibility of a nuclear explosive is beautifully captured in a reminiscence attributed to Edward Teller: That in a March 1939 meeting called to discuss the idea that researchers should begin to withold results from publication, Bohr opined that a nuclear explosive would be impossible "... unless you turn the United States into one huge factory." The issue was the topic of a public debate at an American Physical Society meeting on April 29, with the *New York Times* reporting the next day on the possibility of a nuclear explosion large enough to destroy that city (see Rhodes (1986) pp. 294–297). Bohr's real opinion might have been a little more nuanced, however. In his superb biography of Bohr, Pais (1991) relates that Bohr oscillated between optimism and pessimism through the Fall of 1939, but converged on pessimism late in the year. In a lecture to the Society for the Dissemination of Natural Science in Copenhagen on December 6, Bohr remarked that "...if we had a sufficiently large quantity of the lighter isotope ^{235}U ...every neutron produced in the fission process would have a very considerable probability of causing further fission in a collision with other uranium nuclei, and since for each fission two neutrons are emitted on the average, an explosion would be the unavoidable consequence. With present technical means it is however impossible to purify the rare uranium isotope in sufficient quantities to realise the chain reaction discussed above." The lecture is still worth reading for its clear understanding of the mechanisms of nuclear reactions; a transcript can be found in Bohr's collected works; see Peierls (1986).

Even as Alfred Nier and his collaborators were undertaking their work, Otto Frisch and Rudolf Peierls were considering the very question of a uranium bomb.

The physics stage is now set for a study of the memorandum. But let us first take a qualitative intermission to review the careers of these remarkable men.

References

Anderson, H. L., Fermi, E., & Hanstein, H. B. (1939b). Production of neutrons in uranium bombarded by neutrons. *Physical Review, 55*(8), 797–798

Bohr, N. (1939a). Disintegration of heavy nuclei. *Nature, 143*(3617), 370.

Bohr, N. (1939b). Resonance in uranium and thorium disintegration and the phenomenon of nuclear fission. *Physical Review, 55*(4), 418–419.

Bohr, N., & Wheeler, J. A. (1939). The mechanism of nuclear fission. *Physical Review, 56*(5), 426–450.

Frisch, O. R. (1939). Physical evidence for the division of heavy nuclei under neutron bombardment. *Nature, 143*(3616), 276.

Frisch, O. R. (1973). A walk in the snow. *New Scientist, 60*(877), 833–833.

References

Frisch, O. R. (1978). Lise Meitner, nuclear pioneer. *New Scientist, 80*(1128), 426–428.

Hoddeson, L., Henriksen, P. W., Meade, R. A., & Westfall, C. (1993). *Critical assembly: A technical history of Los Alamos during the Oppenheimer years, 1943–1945*. Cambridge: Cambridge University Press.

Meitner, L., & Frisch, O. R. (1939). Disintegration of uranium by neutrons: a new type of nuclear reaction. *Nature, 143*(3615), 239–240.

Möller, P., Sierk, A. J., Ichikawa, T., Iwamoto, A., & Mumpower, M. (2015). Fission barriers at the end of the chart of the nuclides. *Physical Review C, 91*, Article 034310.

Nier, A. O., Booth, E. T., Dunning, J. R., & Grosse, A. V. (1940). Nuclear fission of separated uranium isotopes. *Physical Review, 57*(6), 546.

Pais, A. (1991). *Niels Bohr's times, in physics, philosophy, and polity*. Oxford: Oxford University Press.

Peierls, R. (Ed.). (1986). *Niels Bohr collected works, Vol. 9: nuclear physics (1929–1952)*. North Holland, Amsterdam

Peierls, R. (1939). Critical conditions in neutron multiplication. *Mathematical Proceedings of the Cambridge Philosophical Society, 35*(4), 610–615.

Reed, B. C. (2014). The Manhattan Project. *Physica Scripta, 89*(10), 108003 (26pp)

Rhodes, R. (1986). *The making of the atomic bomb*. New York: Touchstone.

Szilard, L., & Zinn, W. H. (1939). Instantaneous emission of fast neutrons in the interaction of slow neutrons with uranium. *Physical Review, 55*(8), 799–800.

Turner, L. A. (1939). Nuclear fission. *Reviews of Modern Physics, 12*(1), 1–29.

Chapter 2
Frisch and Peierls

Abstract This chapter offers biographical sketches of Otto Frisch and Rudolf Peierls: Their childhoods, educations, early careers, how they came to be together in Birmingham in 1940, their wartime work, family lives, and postwar careers. The changing political fortunes of the wartime British nuclear program are also discussed. Together, Frisch and Peierls possessed just the right combination of experimental and theoretical backgrounds and skills needed to prepare the memorandum.

It is time to get to know Frisch and Peierls and how their memorandum came to be. The surveys of their lives and careers related here are abstracted largely from their respective memoirs, *What Little I Remember* (Frisch 1979) and *Bird of Passage* (Peierls 1985). Frisch's volume is rather informal, giving only a few details of his accomplishments, but is full of personal reminiscences, interesting anecdotes, and observations on some of the most outstanding personalities of twentieth-century physics. Peierls also relates reminiscences and his interactions with the great figures of modern physics, but goes into much more detail about his own work; both volumes are still worth reading. We start with Frisch, the more senior of the duo.

2.1 Otto Robert Frisch

Otto Robert Frisch was born in Vienna on October 1, 1904, apparently the only child of Justinian and Auguste Frisch. Justinian had studied law but made his living as a painter and then in the publishing business, at which he prospered. Auguste Frisch was a talented concert-level pianist and composer/conductor; she largely gave that up upon marrying, but did teach Otto to play the piano at age five. Originally Auguste Meitner, her younger sister, Lise Meitner–Otto Robert's aunt of fission-discovery fame—was born in 1878.

Otto Robert's aptitude for mathematics was apparent at at a young age; he claims that he could multiply fractions in his head at age five, and at 16 worked out the rudiments of differentials for himself. In 1914, not yet 10, he entered high school. This was a longer curriculum than is now the norm; he graduated in 1922. He then entered the University of Vienna (his aunt's alma mater), majoring in physics and mathematics. The route to a doctoral degree was shorter then; he graduated with that degree in the summer of 1926, not yet 22 years old. His career began with a local instrument-maker that manufactured, among other things, X-ray dosimeters. His break into the wider world of physics occurred in 1927 when he was offered a position in Berlin, likey at the recommendation of his doctoral professor in Vienna, Karl Przibram.

The Berlin position was a three-year post at the Physikalisch Technische Reichsanstalt, a national laboratory akin to a bureau of standards. Lise Meitner was already long-settled in Berlin, and helped her nephew find a place to live while also introducing him to Otto Hahn. He sat in on weekly colloquia at the University of Berlin, which were regularly attended by the likes of Max Planck, Albert Einstein, Walther Nernst, and Gustav Hertz, the pillars of the German physics establishment of the time. He also picked up part-time work at the University, in the laboratory of professor Peter Pringsheim, where he developed an instrument to detect the vapors of potentially hazardous mercury spills which had accumulated over the years in various laboratories.

In 1930, Frisch acquired what he called his first real academic job: As an assistant to Otto Stern, Professor of Physical Chemistry at the University of Hamburg. Stern was a master experimentalist, pioneering the area of atomic and molecular beam research. In collaboration with Walther Gerlach in Frankfurt in 1921/22, Stern conducted an experiment involving passing a stream of silver atoms through a magnetic field; this proved the existence of "space quantization" as predicted by quantum mechanics and led to the discovery of electron spin. Stern would be awarded the 1943 Nobel Prize for Physics for this work.

Reading Frisch's memoir, it is clear that he much loved his time in Hamburg for both the scientific and social environment. Among other experiments, he was involved in measuring the so-called magnetic moment of the proton, showing that beams of hydrogen and helium nuclei exhibited quantum-mechanical diffraction effects, and measuring the speed of recoil of an atom upon emitting a photon.

Unfortunately, this was not to last: The political situation for Jews in Germany was becoming increasing violent; both Frisch and Stern were Jewish. In the summer of 1933, Niels Bohr invited Frisch to a conference in Copenhagen which had been arranged to help displaced physicists find jobs; Bohr would play an enormous role in Frisch's career. That same year, Stern left Germany for America, taking up a position at the Carnegie Institute of Technology in Pittsburgh; this is now Carnegie Mellon University, known for its strengths in engineering and computer science. Before leaving, he arranged for jobs for his assistants, securing Frisch a position in Britain through the Academic Assistance Council (AAC), an organization set up to support academics fleeing Germany. In October of that year Frisch left for London by freighter; it would be years before he set foot in his homeland again.

Frisch's grant from the AAC was good for only a year; it funded him for a position at Birkbeck College in London. Birkbeck is now part of the University of London, but began in 1823 as a college for working men; classes were held in the evening, which gave faculty time during the day to carry out research. The head of the physics department was Patrick Blackett, who had invented the cloud chamber for detecting and photographing reactions induced by cosmic rays; he would be awarded the Nobel Prize for Physics in 1948 for this work. Blackett assigned Frisch the task of attempting to detect the gamma rays which would be created in the mutual annihilation of a positron and an electron. This didn't work out, but upon the discovery of artificial radioactivity in early 1934, he did develop an apparatus to detect the decay products of very short induced half-lives.

Divine providence appeared in the form of Niels Bohr. When Bohr was visiting London, Blackett persuaded him to offer Frisch a position in Copenhagen when the grant ran out. As Frisch wrote in a letter to his mother, 'You need no longer worry about me; God Almighty himself has taken me by my waistcoat button and spoken kindly to me.'

In addition to its remarkable strength in theoretical physics, the Niels Bohr Institue of the University of Copenhagen has a high reputation for experimental work. There, Frisch continued his work on alpha-induced radioactivity, and then shifted to neutron-induced radioactivity following Fermi's pioneering work in that area. The Institute had samples of some rare-earth elements which Fermi did not possess; Frisch and collaborators were the first to explore their neutron-bombardment properties.

Bohr's Institute hosted a constant stream of distinguished visitors, bringing Frisch into contact with the leading personalities of the physics community. One of these was George Placzek, who made cameo appearances in Sects. 1.2 and 1.3; Frisch estimated that Placzek spoke some 10 languages during his time in Copenhagen, and later picked up Hebrew and Arabic as well.

Frisch's immense respect for Bohr as both a physicist and human being shines through in his memoir. For students who are encountering the bewilderments of quantum mechanics, take heart: Frisch quotes Bohr as saying that 'But, but, but ... if anybody says he can think about quantum theory *without* without getting giddy it merely shows that he hasn't understood the first thing about it!' In many ways, nuclear weapons are easier than quantum physics.

In Copenhagen, Frisch was never far from the turmoil in Germany. In March 1938, his native Austria was annexed into Germany; he suddenly found himself a German citizen. When British visitors came through Copenhagen, he would put out feelers for a position back in England.

Frisch's late-1938 involvement in interpretation and verification of fission is related in Sect. 1.2. In his memoir, he describes his visit to Sweden to see Lise Meitner: "This was the most momentous visit of my whole life."

Fortune again smiled upon Frisch in the summer of 1939 in the form of Marcus Oliphant (Fig. 2.1), an Australian native who had been one of Rutherford's many students. Oliphant was head of the physics department at the University of Birmingham, and was one of the British visitors whom Frisch had met in Copenhagen. Igorning bureaucratic formalities, Oliphant invited Frisch over for a summer vacation and appointed him as an auxiliary lecturer to begin in the Fall. Oliphant had taken up his professorship in 1937, and one of his first faculty recruits was Peierls. For a while, Frisch lived with Peierls and his wife, Genia.

The department at Birmingham was deeply engaged in work on radar. As enemy aliens, Frisch and Peierls were barred from such research, but Peierls' knowledge of

Fig. 2.1 Marcus Oliphant (1901–2000). *Source* Public domain, https://commons.wikimedia.org/wiki/File:Sir_Mark_Oliphant.jpg

electromagnetism was too valuable a resource not to utilize. Oliphant circumvented the problem by the guise of posing questions to Peierls in the form of seemingly academic exercises. Both were well aware of the fiction, but it worked. Oliphant would later play a seminal role in prodding American physicists to accelerate their country's fission-bomb efforts.

Frisch's teaching duties were not particularly onerous, and he had free time to pursue his own interests. Aware of Bohr's prediction regarding slow-neutron fission being caused by ^{235}U, he began to contemplate how the theory might be tested. Months before Alfred Nier and his collaborators performed their experiments (Sect. 1.6), Frisch concluded that an approach would be to prepare a sample of uranium in which the proportion of ^{235}U had been artificially increased. If Bohr was correct, then the enriched sample should show an increased rate of fission under slow-neutron bombardment when compared to an unenriched sample. He began to research isotope enrichment methods, and soon zeroed in on the thermal-diffusion method recently developed by two German scientists, Klaus Clusius and Gerhard Dickel. As described in more detail in Sect. 4.9, this involves supplying a gas or fluid of mixed isotopic composition into a heated vertical tube; the fluid containing lighter-isotope molecules will tend to accumulate toward the top of the tube. Frisch had the Birmingham physics department's glassblower prepare a tube; the experiment did not succeed, but his attention soon became drawn in a much more compelling direction.

Unexpectedly, Frisch received an invitation from the Royal Society for Chemistry to write a review article on radioactivity and subatomic phenomena for the 1940 edition of their Annual Report on the Progress of Chemistry. Ironically, he opened his paper with the statement that "The year 1940 has produced no spectacular progress in nuclear physics. The "boom" in papers about nuclear fission . . . has almost faded out." Only a brief mention is made of Bohr's speculation that ^{235}U must be responsible for slow-neutron fission, and the possibility of a chain reaction is raised in one sentence and then dismissed. Frisch later wrote that when he prepared the report, he truly believed that an atomic bomb was impossible. But writing it evidently caused his thoughts to turn back to his enrichment experiment, and he began to wonder if, in the event that he could produce enough pure or highly enriched ^{235}U, would it be possible to make a truly explosive chain reaction based on fast neutrons as opposed to slow ones? Making a rough estimate of the fission cross-section of ^{235}U and using the critical-size formula that had been published by Peierls the previous October, Frisch estimated, to his surprise, a critical mass on the order of a pound.

Frisch's memoirs give the impression that he worked out the critical mass first and then discussed the result with Peierls. On considering the expected efficiency of a single Clusius-Dickel tube, they estimated that a cascade of 100,000 such tubes might be sufficient to produce enough ^{235}U for a bomb in a matter of weeks. As Frisch wrote: "At that point we stared at each other and realized that an atomic bomb might after all be possible." It is at this point that the Frisch-Peierls memorandum began to take shape. As related in the following section, Peierls gives a somewhat different version of the story, but the essence is that they talked the situation over with Oliphant, who forwarded their document to Henry Tizard. As related in Sect. 2.3, this led to the formation of the so-called MAUD committee and the genesis of the British wartime nuclear program.

As their memorandum wound its way through the government bureaucracy, Frisch resumed his work on uranium. Securing some radon gas from the decay of radium which had been used in medical treatments, he constructed a source of slow neutrons. With this, he again verified the fissility of ^{235}U, but realized that he and Peierls had probably underestimated the amount of ^{235}U that would be necessary for a bomb. At about this time, however, Peierls conceived of an offsetting effect, that of surrounding a bomb core with a tamper that would reflect escaping neutrons back into the exploding core and give them fresh chances at causing fissions; this concept would be used in both the Hiroshima and Nagasaki bombs. Frisch also detected spontaneous fission of uranium at about the same time as Flerov and Petrjak in Russia (see Chap. 3).

In mid-1940, Frisch relocated to James' Chadwick's home base, the University of Liverpool, where gaseous uranium hexafluoride would be available for his enrichment experiments; the university also had a cyclotron that could be used as a source of neutrons. His memoir gives an account of German bombing raids, one of which damaged his boarding house and another where a bomb landed in the courtyard of the physics building, blowing out the doors and windows; staff members boarded up the holes and went back to work the next day. This is all related with humour, but the experiences must have been harrowing.

At Liverpool, Frisch continued his work with isotope separation and developing instrumentation for measuring neutron interactions with uranium. Despite his status as an enemy alien, Chadwick trusted him to have extensive contact with researchers in London, Cambridge, and Oxford, where other aspects of the British nuclear program were underway. In particular, Peierls was then at Oxford working with a German-born physical chemist, Franz Simon, on a revolutionary gaseous-diffusion method of enriching uranium that would be put into large-scale use at Oak Ridge, Tennessee; Frisch obtained slightly-enriched samples for his experiments in Liverpool.

In mid-1943, a United States-British negotiation resulted in an agreement to have British scientists travel to America to work at Manhattan Project sites there; this is described in Sect. 2.3. James Chadwick would be the head of the "British Scientific Mission in USA," and, as Frisch relates, approached him one day to ask 'How would you like to work in America?' Frisch replied that he would like that very much, to which Chadwick responded with 'But then you would have to become a British citizen.' Frisch replied that 'I would like that even more.' Within days, he found himself naturalized, equipped with a passport and US visa, and aboard a refitted ocean liner with other native and naturalized Britons enroute to the New World. He recalls that upon his arrival in Newport News, Virginia, one of the first things he saw was well-stocked fruit stands, a sight that sent him into hysterical laughter; in England, he had not seen an orange for over two years.

After a security briefing by the Manhattan Project's Commanding General, Leslie Groves, Frisch traveled by train to his posting at Los Alamos. He was in the first group of British scientists to arrive there, on December 13, 1943. He records that the laboratory director, Robert Oppenheimer, would greet newcomers with 'Welcome to Los Alamos, and who the devil are you?'

In his memoir, Frisch devotes less space to his work at Los Alamos (with a couple exceptions, below) than to the personal memories: Parties at which music was a central theme, skiing and mountain-climbing expeditions, visits to native pueblos, and desert and mountain scenery the like of which he had never previously beheld.

Every new arrival to Los Alamos passed through the office of Mrs. Dorothy McKibbin in Santa Fe, about 35 mi south of the laboratory. Known as the "first lady of Los Alamos," she was one of the first employees of the laboratory, and provided new arrivals with security passes, looked after their families, and generally helped them get settled. She kept an index card on every employee; Frisch's is unusually empty (Fig. 2.2). McKibbin witnessed the Trinity test from around Albuquerque, and remained with the laboratory for 20 years until her retirement in 1963.

Frisch's work at Los Alamos again revolved around instrumentation, but he was also involved in potentially very dangerous experiments known as critical assemblies. These were assemblies of varying amounts of fissile uranium or plutonium built up to approach a critical mass. Short of a real explosion, there was no way to determine the

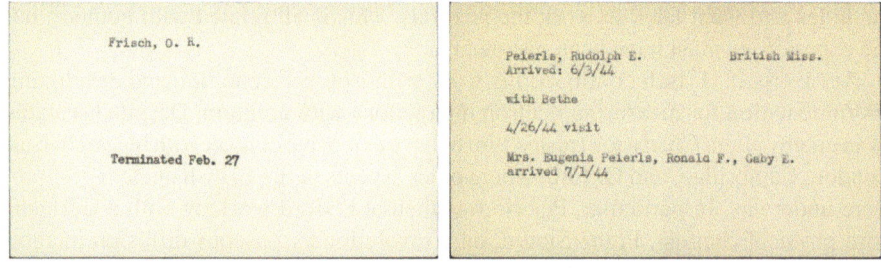

Fig. 2.2 Dorothy McKibbin's file cards for Frisch and Peierls. Peierls' first name is misspelt. *Courtesy* Julie Miller, National security research center, Los Alamos national laboratory

extent of supercriticality that would be achieved with a full-scale bomb, but data from subcritical and barely-critical experiments could be extrapolated to give checks on theoretical estimates. By surrounding a subcritical assembly with neutron-reflective tamper blocks, the number of fissions could be enhanced; such experiments were known as "Godiva" assemblies, where an otherwise bare core would be "clothed" by the tamper blocks. On one occasion, Frisch was assembling tamper blocks while keeping an eye on the flashing red light of a neutron counter. Suddenly, the neutron flux was so great that the light was continuous; the neutron-reflecting effect of his body was making the assembly go critical. He hastily removed some of the blocks, but estimated that if had hesitated for but two seconds more he would have received a fatal dose of radiation. Criticality accidents would result in two postwar fatalities at Los Alamos.

In October 1944, Frisch proposed constructing a device where a slug of enriched uranium would be dropped through the center of an almost-critical assembly of the same material; see Fig. 2.3. When the slug passed through, the assembly would become supercritical for a brief time. Frisch was surprised when a review committee, which included future Nobel laureate Richard Feynman, approved the experiment; Feynman described it as "like tickling the tail of a sleeping dragon," and the experiment became known as the Dragon machine.

The Dragon stood about 6 m high. Designed to be operated largely by remote control, the operator could not activate the "Here We Go" button until various safety interlocks had been activated. A steel box which contained uranium rode on guide wires and was dropped from the top of the device. The box would pass through a lower table on which had been mounted more uranium, producing, for about a

Fig. 2.3 The Dragon machine. Note chair for scale. *Source* Public domain; Malenfant (2005)

hundredth of a second, a very slightly supercritical assembly. Frisch estimated that even if the box became stuck, the resulting explosion would be equivalent to only a few ounces of high explosive.

He was ready by mid-December, and began with tests using dummy materials before moving to active material. On January 20, 1945, the Dragon produced the world's first fast-neutron chain reaction. The reactions were brief, but bursts of up to 10^{15} neutrons were created, accompanied by power releases of up to 20 million Watts and temperature increases in the uranium of up to 2 °C per millisecond over about three milliseconds. There was not a single accident or hangup of material in the drop mechanism. Because other groups needed the uranium, experiments ceased in February, and the machine was subsequently dismantled. Dragon experiments contributed data on such parameters as the generation time between fissions and the exponential growth rate of the chain reaction. A detailed report on the Dragon machine is available in Malenfant (2005).

Frisch witnessed the world's first nuclear explosion, the *Trinity* test of July 16, 1945, from an observation point about 20 mi distant from ground zero. In his memoir, he describes the event:

> And then, without a sound, the sun was shining, or so it looked. The sand hills at the edge of the desert were shimmering in a very bright light, almost colourless and shapeless. The light did not seem to change for a couple of seconds and then began to dim. I turned round, but that object on the horizon which looked like a small sun was still too bright to look at. I kept blinking and trying to take looks, and after another ten seconds or so it had grown and dimmed into something more like a huge oil fire, with a structure that made it look a bit like a strawberry. ... The object, now clearly what has become so well known as the mushroom cloud, ceased to rise but a second mushroom started to grow out from its top; the inner layers of the gas were kept hot by their radioactivity and, being hotter than the rest, broke through the top and rose to even greater height. It was an awesome spectacle; anybody who has ever seen an atomic explosion will never forget it. And all in complete silence; the bang came minutes later, quite loud though I had plugged my ears, and followed by a long rumble like heavy traffic very far away. I can still hear it.

British Mission members were under orders not to get into political discussions; Frisch says little about his opinion of the bombings of Hiroshima and Nagasaki except to remark that while he found the celebratory mood of some of his colleagues rather ghoulish, he recognized that an invasion of Japan would have cost many more casualties. He does add, however, that he and few of his fellow scientists could see any reason for using a second bomb.

Following the end of the war, Frisch spent a brief holiday in California, but then received an unexpected offer. A contingent of British scientists worked on reactor theory and development in Montréal, Canada. This group, under the direction of John Cockcroft, another protégeé of Ernest Rutherford, developed the first reactor to go critical outside the United States, the so-called Zero-Energy Experimental Pile; criticality was achieved on September 5, 1945. The British government was moving to establish its own nuclear research facility, the Atomic Energy Research Establishment in Harwell, outside Oxford. Cockcroft recruited Frisch to be a Division Leader in the new facility, with the title of Deputy Chief Scientific Officer. Frisch spent a few weeks in Canada before heading back to Britain. One of his colleagues there was an acquaintance from Los Alamos, Klaus Fuchs, who would later be unmasked as a spy who provided information to Russia; Frisch was astounded when Fuchs was charged with treason in 1950. After a quick trial, Fuchs was sentenced to 14 years in prison, but served only nine; he emigrated to East Germany after being released. Fuchs' motive was apparently not money but rather to help a wartime ally; Frisch was convinced that Fuchs was sincere in his beliefs.

At Harwell, Frisch was mostly involved with the theory of chain reactions and their fluctuations. However, in the spring of 1947, he was offered Ernest Rutherford's professorial position at Cambridge University, an opportunity too attractive to pass up. He would remain there for the rest of his career until his mandatory retirement at age 67.

Through an acquaintance in London, Frisch met Ursula (Ulla) Blau, a fellow Viennese expatriate and skilled graphic artist and author of children's books. They married in the spring of 1951, and had two children, Monica Eleanor and David Anthony (Tony). Tony followed in his father footsteps by becoming a physicist, and Monica went into a career in the social sciences. Ulla outlived Robert by 40 years, passing away in Cambridge in September 2019 at the age of 98. A brief biography of Ulla by Monica can be found at https://www.geni.com/people/Ursula-Ulla-Frisch/6000000084813969937.

Frisch's work at Cambridge returned to instrumentation, in particular devising automated means of analyzing bubble-chamber photographs obtained in particle-physics experiments. His thoughts on the development of computers are still striking: "... I am convinced that most of us are nowhere near to understanding their full importance. Two hundred years from now historians will say that the computer changed our world much as the steam engine did two hundred years previously, if not more so."

With Frisch's passing on September 22, 1979, the world lost one of the pioneers of the age of nuclear fission who had been present at its creation.

2.2 Rudolf Ernst Peierls

Rudolf Peierls (most commonly pronounced "pyerls" or "puyerls", akin to "buy") was born in suburban Berlin on June 5, 1907, the youngest child of Heinrich and Elisabeth Peierls, a sibling to Alfred and Annie. The family was reasonably well-off; Heinrich Peierls was the managing director of a large factory of the Allgemeine Elektrizitäts-Gesellschaft, a manufacturer of electrical cables. As with the Frisch family, music was an important element of social occasions.

Peierls did not find school particularly challenging, but benefitted from excellent mathematics and English teachers; he also read widely in the popular sciences. The family sufferd some privations during World War I, but his father's position guaranteed a fairly stable life even during the rampant inflation that prevailed in Germany in the 1920's. Tragically, his mother died when he was 14; not long thereafter, his father married their housekeeper.

Peierls entered the University of Berlin in the winter semester of 1925 to study physics. In those days, there was no set syllabus of courses or even the equivalent of an undergraduate degree; the first examination one would pass was for the Ph.D. In his first semester, Peierls was registered for 36 h of lectures per week, concentrating heavily on theoretical and mathematical classes. One of the required courses was philosophy, of which he had a low opinion, agreeing with the definition that "Philosophy is the constant misuse of a terminology specially invented for the purpose." He also relates that the lectures of quantum pioneer Max Planck were particularly bad, with Planck reading verbatim from his own books.

After two semesters in Berlin, Peierls transfered to the University of Munich, home of the renowned theoretical physicist Arnold Sommerfeld, who was making seminal contributions to the newly-developing area of quantum mechanics. Another student there and a year older was Hans Bethe. The two struck up a lifelong frienship, and would later live together in Manchester, England. Bethe would emigrate to America in 1935 to take up a position at Cornell University, and they would reunite at Los Alamos, where Bethe headed the Theoretical Division.

In the spring of 1928, Sommerfeld left for a lecture tour in America for a year, and arranged for Peierls to move to Leipzig to study with one of his former students, Werner Heisenberg—yes, *that* Heisenberg of Uncertainty Principle fame. As well as describing the theoretical problems Heisenberg had him investigate (electron spin, spectral lines, and conductive effects in metals), Peierls comments in his memoirs on Heisenberg's intense competitiveness when playing table tennis. A visitor to Leipzig whom Peierls got to know was George Placzek, who appeared in Chap. 1.

In the spring of 1929, Peierls moved again when Heisenberg also went to America for a while; this time he shifted to Zurich to work with Wolfgang Pauli at the Eidgenössische Technische Hochschule, the German-speaking Swiss Federal Institute of Technology. There he took up an analysis of vibrational phenomena of atoms in crystals, work which became his thesis topic that summer. To satisfy a residency requirement the thesis was submitted to Leipzig, to which he had to return temporarily for the defense examination. He was now "Herr Doktor" Peierls.

Before Peierls left for Leipzig, Pauli offered him the position of being his assistant for a year, what now would be considered a post-doctoral appointment. This was an honor, as Pauli had a reputation of being quite scathing about other's abilities. Many sarcastic quotes are attributed to him, such as "You know, what Mr. Einstein said is not so stupid!", "I do not mind if you think slowly, but I do object when you publish more quickly than you can think", "What you said was so confused, one could not tell if it was nonsense!", and "This isn't right. This isn't even wrong." The assistantship would stretch into three years.

Peierls returned to his work on the theory of electron conduction in metals. During his 1930 Easter vacation he visited Copenhagen to meet Niels Bohr, the first of what would be many visits. The two remained close until Bohr's passing in 1962.

In the summer of 1930, Peierls was invited to attend a conference in Odessa on the Black Sea, then in the Soviet Union but now in Ukrainian territory. There he met Eugenia (Genia) Kannegiser, a recent physics graduate from Leningrad; they hit it off and married the following spring (Fig. 2.4). Peierls returned to Zurich, but kept

Fig. 2.4 Genia and Rudolf Peierls in New York, 1943. *Source* Available for unrestricted use. Photograph by Francis Simon, *Courtesy* of AIP Emilio Segrè Visual Archives, Simon Collection

in touch with her by letter; they did not see each other again until he could return to Leningrad at the end of the winter semester in 1931. They married on March 15, but had only three weeks together before he had to return to Zurich for the summer; she could not go with him as she had to apply for permission to change citizenship. He returned at the end of the summer semester, by which time her permission had come through. Once in Germany, she automatically become a German citizen by virtue of the marriage.

By mid-1932, Peierls' time as Pauli's assistant was coming to an end. Pauli supported Peierls' application for a Rockefeller Foundation fellowship, which would support him to study abroad for a year. A condition of the fellowship was that an applicant had to have a position to return to; to cover this, Pauli claimed that Peierls could return as his assistant, but with the mutual understanding that he would not. The tenure of the fellowship included part of 1933, during which the world would change dramatically.

Peierls decided to split his fellowship between working with Enrico Fermi in Rome for six months and then moving to Cambridge. The Peierls' arrived in Rome in October just when Fermi was beginning his historic experiments with neutron bombardment; Peierls did not participate in these, but instead learned techniques for numerically solving differential equations, a skill that would come in handy at Los Alamos.

By Christmas of 1932, Genia was pregnant with their first child, and an unexpected offer arose: A position in Hamburg to begin at Easter 1933. Peierls was attracted to returning to his native country, but they decided against the offer given the increasingly violent political situation. In the spring of 1933, they departed Rome for Cambridge.

As the home of Ernest Rutherford, James Chadwick, and many other distinguished experimental and theoretical physicists, Cambridge University was one of the leading physics research institutions in the world. At the time, a laboratory devoted to studies in magnetism was being established, an excellent match with Peierls' interests in the phenomena of electron transport in metals. In August 1933, the Peierls' first child, daughter Gaby, was born.

Peierls' situation was uncertain: His Rockefeller Foundation fellowship would expire in the fall, and positions were getting harder to come by as refugee scientists were fleeing Europe. They certainly could not return to Germany. At Manchester University, William Bragg, who had shared the 1915 Nobel Prize in Physics with his father for discoveries in X-ray diffraction, arranged for Peierls to be supported by a two-year grant from the Academic Assistance Council, the same organization that had brought Otto Frisch to Britain. The family would be making yet another move.

Hans Bethe was then in Manchester, and he lived with the Peierls' for a year to share expenses; the two began collaborating on papers in nuclear physics. It was not long, however, until there were visits back to the continent. In the fall of 1933 Peierls was invited to attend that year's Solvay Conference in Brussels, a very prestigious gathering; other attendees included Bohr, Chadwick, Fermi, Heisenberg, Rutherford, and Lise Meitner. In the spring of 1934 he was invited to lecture at the Institue Henri Poincaré in Paris, and that summer they visited Genia's parents in Leningrad.

An unending concern, however, was securing more permanent employment. Peierls considered applying for positions in places as far-flung as India and Ecuador, but in the spring of 1935 he was offered a temporary appointment back at the magnetism laboratory in Cambridge. Their second child, son Ronald, was born shortly before they left Manchester.

In 1936, Marcus Oliphant, whom we met earlier as Otto Frisch's savior, was appointed head of the physics department at Birmingham University. He delayed taking up the post until the fall of 1937, but convinced the university that it needed to establish a professorship in applied mathematics, and sounded Peierls out on the position. A formal interview process was held; Peierls was offered the position, which he eagerly accepted as it was permanent and carried a salary 2.5 times what he was making at Cambridge; the Peierls' celebrated by buying a used car for 25 pounds. Peierls initiated a new program in mathematical physics; it was not uncommon for British universities to offer no courses in quantum mechanics until into the late 1930's.

The Peierls' would remain in Birmingham for five years until war work took them to America. Stimulated by his work with Hans Bethe, Peierls begin working on the theory of nuclear reactions, unkowingly preparing himself for what was to come. In 1938, he made several trips to Copenhagen to work with Bohr and Placzek on the theory of neutron-nuclei collisions, a matter that would soon become of paramount importance. The threat of war was always in the air, however, and Peierls kept in touch with his old friend Hans Bethe, indicating that he would be interested in any job in the United States.

When the war broke out, the Peierls' found themselves to be enemy aliens. Fortunately, he had begun the process of applying for naturalization, which came through in February 1940. He was still not allowed to work on Oliphant's radar project, but did volunteer with an Auxiliary Fire Service brigade which helped fight fires started by air raids, and Genia worked as an auxiliary nurse. In the summer of 1940, they took advantage of an offer from the University of Toronto to care for the children of academic staff at Birmingham and Oxford. Gaby and Ronald, then six and four, made the trip, two of thousands of British children who were sent to safety overseas. At this point, Otto Frisch moved in with the Peierls' until his own move to Liverpool.

We come now to Peierls' version of the genesis of the memorandum.

As related in Chap. 1, Peierls read Francis Perrin's paper on criticality, and realized that he could refine the calculation. At that time, he did not substitute any numbers into his expressions; his paper was entirely analytic. One cannot help but wonder if he would have published had he been in possession of even approximately correct values for the relevant cross-sections for ^{235}U.

As Peierls described the situation in an April 1985 interview with historian Mark Walker, " ... I wondered whether it was sensible or reasonable to publish this paper because it might have some connection with a weapon and maybe there was something I shouldn't publish about it. So I talked to Frisch, who was in Birmingham at this time, and between us we decided that since Bohr had given a reasonable assessment that you could not make a bomb with uranium, thinking of natural uranium, there was no question of a weapon and therefore it was perfectly safe to publish this paper. And it was only then when Frisch came along and said, "Now, supposing somebody gave you a large quantity of separated uranium, what would happen?" Then (with my paper, my formula, one could make a reasonably good guess at the cross-section from Bohr's theory) we came out of it with an estimate of critical size which was surprisingly small because one had thought about tons. And then we said, "All right, now let's see what will happen if you get such a chain reaction" and since I also had some estimates on the time scale, and it was a question of the competition between the expansion of the material over the developing heat and the chain reaction, and we concluded that you would get a high efficiency, we didn't know how high, but it was clearly a tremendous value." The full interview can be found at https://www.aip.org/history-programs/niels-bohr-library/oral-histories/4819.

In his memoir, Peierls goes on to say that states that they went on to estimate how much energy the reaction might liberate until the uranium dispersed itself. The result was equivalent to thousands of tons of ordinary explosive. He also relates that, in a classic understatement, they said to themselves: "Even if this (diffusion) plant costs as much as a battleship, it would be worth having."

Alarmed at the idea that German scientists might be thinking along the same lines, they felt it their duty to inform the British government of the possibility of atomic weapons, but in a way that would keep their work secret in case German researchers hadn't yet thought of it (they had). They decided to prepare a memorandum on the matter, which Peierls typed up himself rather than entrust to a secretary. They kept only one carbon copy.

At the time, Peierls was newly naturalized and Frisch was still an enemy alien; they were not sure how to get their ideas to appropriate officials. They took their documents to Oliphant, who forwarded them Henry Tizard. In a cover letter, Oliphant related that "I have considered these suggestions in some detail and have had considerable discussion with the authors, with the result that I am convinced that the whole thing must be taken rather seriously, if only to make sure that the other side are not occupied in the production of such a bomb at the present time. In fact, I view the matter so seriously that immediate steps should be taken to connect with the necessary authorities concerning the possibilities of palliative measures if such a bomb should be used." British historian Ronald Clark found the non-technical memorandum among Tizard's papers some years later, and deduced that the documents reached him on March 19, 1940. In another confluence of fission-history events, this was just four days after the publication date of the Nier-Columbia verification that ^{235}U is responsible for slow-neutron fission. As related by Lorna Arnold, the technical appendix was found in a cornflake box in a storage area of the UK Atomic Energy Authority in the early 1960's. The so-called MAUD committee that was set up in response to the memorandum is described in the following section; for now, we return to Peierls.

With the realization that the key to a bomb would be securing enriched uranium, Frisch and Peierls turned to the question of enrichment techniques, focusing on the method of gaseous diffusion, an alternative to the thermal diffusion method discussed in the section on Frisch above. The fundamental principle of gaseous diffusion is that if a gas of mixed isotopic composition is pumped against a porous barrier containing millions of microscopic holes, atoms of lower mass will on average pass through slightly more frequently than those of higher mass. The result is a vey minute level of enrichment of the gas in the lighter-isotope component on the other side of the barrier. Since only a small enrichment factor can be achieved in any one step, the slightly-enriched gas has to be pumped on to subsequent enriching stages. By linking together a number of processing cells in series in a cascade, bomb-grade material can eventually be isolated. The gas which emerges from each stage slightly "depleted" in the lighter isotope does however still contains atoms of that isotope and so needs to be recycled to an earlier stage for additional processing. While this sounds simple in principle, there are many complications such as getting holes of the correct size, finding a gas that will not be corrosive, and predicting the effects of operating disruptions on the cascade. Peierls devoted considerable time to this effort, and in May 1941 brought in another refugee theoretician, Klaus Fuchs, to help with the work—a hire which, as is now well-known, would have momentous consequences. Peierls and Fuchs would grow especially close, with Fuchs practically becoming a member of the family. Their relationship and the work of security services in eventually revealing Fuchs' treachery are compellingly related in Close (2020).

With the summer 1941 conclusion of the MAUD committee that a nuclear weapon would be feasible, the pace of work picked up, although it was clear that the necessary industrial plants would likely not be able to be built in wartime Britain. In late 1941, Peierls was part of a delegation that visited various laboratories in the United States; during this visit he met with several leaders of the American nuclear program, including Enrico Fermi, Arthur Compton, and Robert Oppenheimer. He was also able to visit his father and sister in New York, and Ronald and Gaby in Toronto.

As described in the following section, relations between Britain and America on nuclear matters soured through 1942 before resuming in mid-1943. Peierls kept at his theoretical work, refining numerical techniques for solving differential equations.

With the (limited) resumption of Anglo-American nuclear cooperation following the signing of the Quebec agreement in August 1943, Peierls, Oliphant, Franz Simon, and Chadwick were the first group to visit Washington under the new arrangement. This was a brief visit for Peierls, during which he met General Groves and was briefed on security arrangements. He returned to Britain, but would soon find himself be back in America as a member of the British Mission.

The Peierls' made the transatlantic crossing on the same ship as Otto Frisch. Frisch went directly to Los Alamos, but the Peierls' stayed in New York for several months so that he could consult on the design of the enormous gaseous diffusion plant being built in Tennessee. In the summer of 1944, they moved (with the children) to Los Alamos, where Peierls was assigned to work on implosion calculations relevant to the plutonium bomb in the Theoretical Division, which was headed by Hans Bethe; one of his colleagues in that work was again Klaus Fuchs. Chadwick appointed Peierls as head of the British group in Los Alamos; Chadwick himself spent most of his time in Washington. Peierls and Robert Christy, a student of Robert Oppenheimer, jointly patented the implosion design for the Trinity bomb; see Chadwick and Chadwick (2021).

Peierls witnessed the Trinity test from Campaña Hill, the same observation point as Frisch. As he writes in his memoir:

> The big moment came: a giant flash, and a fireball rising and turning into the by-now-familiar mushroom-shaped cloud. We were struck with awe. We had known what to expect, but no amount of imagination could have given us a taste of the real thing.

As to the decision to use the bomb and the idea of of a demonstration shot, Peierls offered this opinion:

> To me the obvious answer would have been to drop a bomb on a sparsely populated area to show its effects, coupled with an ultimatum to the Japanese government to negotiate for peace to avoid a large-scale nuclear attack. This would have involved killing some people and destroying some buildings, since otherwise the power of the bomb would not have been obvious; the effects visible after the Alamogordo test were frightening to the expert but not impressive to the layman. Of course such an ultimatum might have failed, but at least it would have been an attempt to avoid unnecessary casualties. ... My regrets are that we did not insist on more dialogue with the military and political leaders, based on full and clear scientific discussions of the consequences of possible courses of action. It is not clear, of course, that such discussions would have made any difference in the end.

In July 1945, Peierls prepared a report on the early work of the British "Tube Alloys" program; this can be found in Moore (2021).

The Peierls' returned to Britain after the war. Rudolf had been offered a position at Cambridge, but decided to return to Birmingham to build up theoretical physics there. He would remain at Birmingham for 17 years, his longest tenure in one place. The Peierls' third and fourth children, Catherine (Kitty) and Joanna (Jo) were born there in 1948 and 1949, respectively. The children ended up in diverse careers: Gaby attended Oxford and went into investment analysis; Ronald took after his father and studied theoretical physics at Cambridge and thereafter became a grad student of Hans Bethe's at Cornell; Kitty became a high-level medical technician, and Jo a computer programmer. Several grandchildren also appeared.

Like Otto Frisch, Peierls was shocked by Klaus Fuchs' arrest for espionage, and visited him in prison. Since it was Peierls who had brought Fuchs into the wartime program, he too came under suspicion, and this led to several unpleasant episodes with unscrupulous tabloid journalists publishing fabricated quotes attributed to both he and Genia.

At Birmingham, Peierls resumed research in quantum electrodynamics, nuclear physics, and solid-state physics, working with many students and collaborators who went on to distinguished careers of their own. Under his influence, theoretical physics became much stronger at the university, attracting students from around the world. Despite his administrative, teaching, and research responsibilities, he maintained an active travel schedule of lectures, conferences, and summer schools in locales including France, Mexico, Canada, Italy, Russia, the United States, and Japan; one notable visit to Trieste, Italy, in 1972 gave him the opportunity to renew acquaintance with Werner Heisenberg. He also served as a consultant to the UK's Atomic Energy Research Establishment and the nuclear power division of the English Electric Company; he was also active in the British Atomic Scientists Association, a group devoted to educating the public on atomic energy and nuclear disarmament. This organization petered out in the 1950's, but he remained active in the much larger internationally-based Pugwash disarmament movement, appalled with the excessive stockpiles of weapons built up by the various nuclear powers.

In 1961, Peierls was offered the position of Wykeham Chair of Physics at Oxford University, the chairmanship of the university's department of theoretical physics. As a condition of taking the position, he insisted that new positions and facilities be funded. He accepted the position in early 1962, but delayed his move until the fall of 1963 to wrap up his work at Birmingham. As at Birmingham, theoretical physics expanded and flourished under his leadership.

In 1967, the Peierls' spent a half-year sabbatical at the University of Washington in Seattle, and became so enamored of the area that they would visit for a month or two each year. The 1967 visit gave them an opportunity to call on Robert Oppenheimer shortly before his death. Peierls was knighted in 1968; they were now to be addressed as Sir Rudolf and Lady Genia; he remarked that one benefit of this was that "Sir Rudolf" was easier for people who struggled to pronounce his surname.

In 1974, Peierls reached the compulsory retirement age at Oxford, although he remained on as an emeritus professor and continued to teach. A retirement symposium was attended by some 200 guests, including Hans Bethe.

Post-retirement, the Peierls' kept up their usual vigorous travel pattern with a succession of visiting professorships and lectureships in Europe, North and South America, and Asia. During an extended visit to Australia, they were hosted by their old friend Marcus Oliphant, then living in Adelaide and serving as the Governor of South Australia. In 1983, he met many old acquaintances at Los Alamos during a 40th anniversary celebration of the founding of the laboratory. Among Peierls' many awards an honors, one that stands out is the 1980 Enrico Fermi Award given by the United States Department of Energy for lifetime achievement in the development, use, or production of energy.

Genia Peierls passed away in October 1986, a year after the publication of Rudolf's memoir. He passed away at Oxford on September 19, 1995, just over 50 years after Hiroshima and Nagasaki.

2.3 MAUD, Roosevelt, and Churchill: Bomb Politics

This section is largely adopted from my *The History and Sciences of the Manhattan Project*, which relies heavily on the volumes of Gowing (1964), Clark (1961, 1965), Farmelo (2013) and Ruane (2016). A a recent treatment of the political machinations behind the Anglo-American wartime and postwar nuclear relationships can be found in Lee (2022), and excellent surveys of the British Mission at Los Alamos can be found in Fakley (1983) and Szasz (1992).

As described above, Marcus Oliphant forwarded Frisch and Peierls' memorandum to Henry Tizard. Tizard was initially skeptical that any practical bomb could be made with uranium, but had to take the possibility seriously.

When he recieved the memorandum, Tizard had already been in contact regarding fission with George P. Thomson, professor of physics at Imperial College, London and son of J. J. Thomson of electron-discovery fame. When Hans von Halban and his collaborators had published their discovery of approximately three neutrons emitted per fissioning uranium nucleus (April 1939), Thomson began to consider the possibility of achieving a chain reaction if a sufficient mass of uranium could be brought together. He assembled a pile with uranium oxide and tried water and paraffin as moderators, but did not achieved a chain reaction. When he was contacted by Tizard, he had almost come to the conclusion that atomic energy was not worth pursuing as a war effort.

James Chadwick initially also doubted the idea of a uranium bomb, but began to reconsider with the appearance of the Bohr-Wheeler fission theory in September 1939. In October, he was contacted by Professor Edward Appleton, Secretary of the Department of Industrial and Applied Research, who inquired whether he thought the possibility of a uranium bomb merited concern. Chadwick reported back in early December that, while could give no definite answer, he would initiate experimental work, and began readying his cyclotron at the University of Liverpool to make measurements of the fission cross-section of uranium for fast neutron bombardment. Privately, Chadwick expressed concern to colleagues that British laboratories seemed disorganized, and feared that leadership in physics would shift to the United States, exactly what would come to pass.

It was against this background that that the Frisch-Peierls memorandum reached Tizard, who prevailed upon Thomson to convene a committee to investigate the matter. Thomson served as chair; the members included, among others, Chadwick and Oliphant. Ironically, Frisch and Peierls, being refugees, were barred from serving on the committee, and so were initially excluded from learning what became of their work. Frustrated, they sent Thomson a ten-page memo on the "uranium problem" in late July; Thomson arranged a compromise whereby they could serve as consultants. When the work of the committee was split into two groups in March 1941, a Policy Committee and a Technical Committee, Frisch and Peierls were allowed to serve on the latter.

Thomson's group came to be called the MAUD Committee. This unusual name had a curious provenance. In April 1940, German forces occupied Denmark. As this was happening, Niels Bohr sent a telegram to Otto Frisch through Lise Meitner, the six concluding words of which were "Tell Cockcroft and Maud Ray Kent." Cockcroft was John Cockcroft of Cambridge University, but the meaning of "Maud Ray Kent" was a mystery. One theory was that by changing the "y" to an "i," Maud Ray Kent became a quasi-anagram for "radium taken." Another rearrangement of letters was interpreted as a plea to enrich uranium: "make ur day nt." Somebody suggested MAUD as a cover name for the committee, and the appellation stuck. Officially, it had periods between the letters (M.A.U.D.), but I will use the simplified form. The mystery was not resolved until after the war. Maud Ray lived in Kent, and had at one time served as a governess for Bohr's children. Her address was to have appeared between "Ray" and "Kent," but was lost in transmission.

The committee held its first meeting on April 10, 1940; within weeks, the Battle of Britain would be in full engagement. Thomson began to take the idea of a bomb seriously, and on the 16th wrote to Chadwick to say that the concept "is not so impossible when you come to look into it." By the summer of 1940, research was underway at the universities of Liverpool (cross-section measurements), Birmingham (uranium chemistry), Cambridge and Oxford (separation methods), and at Imperial Chemical Industries. Peierls spent the summer studying isotope separation methods, and reported in September that the most promising approach looked to be gaseous diffusion through a mesh of fine holes; experiments were already underway by Franz Simon at Oxford. By December, Simon's group was far enough along to estimate parameters for an actual production plant. For an output of 1 kg of ^{235}U per day, some 70,000 square meters (17 acres) of diffusion membrane would be required; the plant would cover some 40 acres, and consume power at a rate of about 60 megawatts. Simon's estimates of the cost of plant construction and the number of operators required would prove far too low, but the important thing was that thoughts on atomic bombs were moving toward practical engineering considerations.

At the same time as enrichment techniques were being considered, Chadwick's cross-section measurements were tending toward confirming Frisch and Peierls' theoretical analysis, and his initial skepticism began to turn to worry. From a 1969 interview conducted at the American Institute of Physics: "I remember the spring of 1941 ... I realized then that a nuclear bomb was not only possible–it was inevitable. ... I had many sleepless nights. ... And I had then to start taking sleeping pills. It was the only remedy. I've never stopped since then. It's been 28 years, and I don't think I've missed a single night in all those 28 years." (Chadwick 1969)

By March 1941, Peierls was convinced that a bomb was distinctly possible. As quoted by Gowing (1964, p. 68), he wrote that "there is no doubt that the whole scheme is feasible ... and that the critical size for a U sphere is manageable." On April 9, he reported his conclusion to a meeting of the MAUD committee, which in early summer began to prepare its final report.

In America, the equivalent group was the "National Academy of Sciences Committee on Atomic Fission" chaired by Nobel laureate Arthur Compton of the University of Chicago. This group was established in the spring of 1941 to look into possible military aspects of fission. In three reports prepared in May, June, and November of that year, the committee outlined the possibilities of power production (reactors)

2.3 MAUD, Roosevelt, and Churchill: Bomb Politics

and bombs, although its initial emphasis was on the former. Compton's reports went to Vannevar Bush and James Conant, who headed the Uranium Committee of the United States' National Defense Research Committee (NDRC), which was tasked with coordinating research that might be of military value. Bush was a professor of Electrical Engineering at the Massachusetts Institute of Technology, and Conant a distinguished chemist and President of Harvard University. In June 1941 the NDRC became the Office of Scientific Research and Development (OSRD). The Compton committee's November report would be significantly if unofficially influenced by the July 1941 MAUD report.

The success of the Manhattan Project under U. S. Army leadership and the fact that the bulk of its facilities were located on American soil have tended to cast the Project as an almost exclusively American affair. But such a view trivializes very important British contributions to the Project. Even General Groves, who in his memoir *Now It Can Be Told* [Groves (1983)] described the British contribution as "helpful but not vital" (p. 408), tempered his assessment with the observation that "I cannot escape the feeling that without active and continuing British interest there probably would have been no atomic bomb to drop on Hiroshima. The British realized from the start what the implications of the work would be. They realized that they must be in a position to capitalize upon it if they were to survive ... and they must also have realized that by themselves they were unable to do the job. They saw in the United States a means of accomplishing their purpose."

American authorities were not unaware of progress in Britain; exchanges between the two countries on scientific matters were well-established before America entered the war. In late August 1940, a mission headed by Tizard left for a two-month visit to America, where they demonstrated progress that had been made with equipment relating to radar and proximity fuses. One of the results of this visit was the establishment in Washington of a formal organization to facilitate information exchange, the British Commonwealth Scientific Office. In the spring of 1941, Charles G. Darwin– a grandson of *the* Charles Darwin–was appointed as its Director. Reciprocally, in February 1941, James Conant traveled to London to set up an office of the NDRC; he also met with Winston Churchill. Frederick Lindemann, Churchill's personal science advisor, spoke with Conant regarding the Frisch-Peierls memorandum; Lindemann's role in the British nuclear project is discussed later in this section. Harvard physicist Kenneth Bainbridge, who would direct the *Trinity* test, attended the April 9 meeting of the MAUD committee at which Rudolf Peierls reported that a fast-neutron bomb was feasible, and on July 1, California Institute of Technology physicist Charles Lauritsen attended another meeting at which the main conclusions of the committee's report were discussed. Lindemann was also present at this latter meeting, and was briefed by Chadwick and Peierls. Lauritsen returned to the United States and briefed Bush in Washington on July 10, just a few days before he received a draft copy of the report which had been transmitted to the NDRC office in London.

There were actually two MAUD reports, both authorized by Thomson on July 15, four years plus one day before the *Trinity* test. The first, which is the one of interest here, was titled "Use of Uranium for a Bomb"; the second was "Use of Uranium as a Source of Power." Both are reproduced in Margaret Gowing's book on the British atomic energy program, and are still worth reading. The first part of the bomb report summarizes the situation in non-technical terms in a few pages. It opened with a description of why a critical mass exists for a fissile isotope, how a bomb could be triggered by bringing together two subcritical masses, the probable effects of the explosion (estimated as equivalent to 1800 tons of TNT for 25 pounds of ^{235}U), and a discussion of materials and costs. A technical appendix described how a fast-neutron chain reaction cannot be sustained in ^{238}U due to the presence of inelastic scattering and neutron capture, how the efficiency of a bomb could be estimated, factors that affect the determination of critical mass, estimates of damage, and the characteristics of a diffusion plant. Depending on values adopted for cross-sections, secondary neutron numbers, and whether or not a bomb was tamped, the critical mass was estimated to lie between about 2 to 43 kg; the latter figure is remarkably close to the presently-accepted value for an untamped ^{235}U core.

The overall conclusion of the report was that a uranium bomb was possible and likely to lead to decisive results in the war, and urged the government to pursue the project as a matter of high priority, predicting that it could be carried out in about two and a half years. Chadwick and Lindemann favored a strictly British effort, but Tizard felt that Britain should collaborate with the United States.

Thomson handed Bush and Conant copies of the report on October 3, and also met with the Compton Committee to apprise it of the situation. Also, during August and September 1941, Marcus Oliphant traveled around the United States, speaking with various physicists about the project and encouraging them to take action. Oliphant visited Berkeley and met with Ernest Lawrence, who was so impressed with British progress that he began thinking of how he might turn his 37-inch cyclotron into a large-scale mass spectrometer for separating isotopes. Oliphant would return to America in 1943 to work on this electromagnetic method of isotope separation, splitting his time between Berkeley and Oak Ridge, Tennessee. An excellent summary of Oliphant's life can be found in an obituary by Dalitz (2001).

Upon returning to Britain, Oliphant was horrified to learn that the government had decided to turn the MAUD Committee over to Imperial Chemical Industries, which saw lucrative possibilities in the postwar energy field. ICI's effort would be headed by the company's research director, Wallace Akers, a very competent and diplomatic industrial chemist. However, MAUD scientists felt that they would be coming under the leadership of commercial representatives who were ignorant of the physics involved. Oliphant resigned from the committee in protest, although he later conceded Akers' competence. It was at this point that the British program became code-named "Tube Alloys."

Despite its early start with the Frisch-Peierls memorandum and the MAUD committee, the British program suffered from serious political mishandling. Frederick Lindemann, Churchill's scientific advisor, was an Oxford physics professor. He had taken up his position at Oxford in 1919; while active in aeronautical research during World War I, he gave that up to be more of a popularizer of scientific developments. Moving comfortably in high British social circles, he and Churchill first met in the early 1920's, and they developed a strong friendship. When Churchill became First Lord of the Admiralty (head of the Navy) at the outbreak of World War II, he

2.3 MAUD, Roosevelt, and Churchill: Bomb Politics

appointed Lindemann as his private scientific advisor, and when Churchill became Prime Minister in May 1940, Lindemann became one of the most influential scientists ever to serve in government. Over the course of the war, he forwarded some 2,000 briefing papers to Churchill.

However, Lindemann's position enabled him to sideline advice from more informed sources such as Tizard, who resigned from the Air Ministry in frustration in the summer of 1940. In October, Churchill moved to blunt some of the criticism of Lindemann by the bureaucratic maneuver of appointing a Scientific Advisory Committee under the chairmanship of senior civil servant Lord Maurice Hankey, but Lindemann's influence was so strong that it would take some time for Churchill to appreciate the revolution in strategic thinking that nuclear weapons portended. Churchill dismissed Hankey in March 1942 when he questioned Lindemann's influence one too many times.

The formal route of the MAUD report was to be to go through hearings in the Scientific Advisory Committee, but Lindemann was not about to wait for that. In a six-page memo dated August 27, 1941, he apprised Churchill of the contents of the report. Lindemann stated that while it seemed almost certain that a bomb could be made, he was skeptical of the two-year timeline, giving it odds of no better than even. He did however advise that the project go ahead on the grounds that if the Germans were to acquire such a weapon, they could defeat England or reverse the verdict of the war after England had defeated them, and also that Britain should undertake the work on its own. Less than two years later, the British would be swimming against a similar exclusionary perspective from the other side of the Atlantic. In a memo to his Chiefs of Staff on August 30, Churchill urged that no expense be spared to push the project, thus becoming the world's first national leader to sanction a nuclear weapons development program. On September 3, the Chiefs endorsed the project, advocating that the development should take place in Britain, and concurring with Churchill's recommendation that the project be overseen by Sir John Anderson, a member of the War Cabinet and an extremely competent administrator. Lindemann wanted no part in the actual mechanics of overseeing the undertaking; if a bomb proved impractical, he did want to be associated with an expensive failure.

Anderson's involvement in wartime British administration ran deep: He would also serve as Chancellor of the Exchequer from September 1943 to July 1945; he has been described as the glue that held the British nuclear program together. In one important way, however, his opinion on the project differed from that of Churchill and Lindemann, which put him in an awkward position: He was firmly of the opinion that the bomb should be built in America. Lindemann briefed the Scientific Advisory Committee on September 17, but in their own report of a week later that group advocated that a gaseous diffusion plant be built in Canada.

Less than a month later, on October 12, a golden opportunity landed on Churchill's desk in the form of a private letter from President Roosevelt, who suggested that the two leaders "correspond or converse concerning the subject which is under study by your MAUD Committee and by Dr. Bush's organization in this country, in order that any extended efforts may be coordinated or even jointly conducted . . ." Churchill, however, was wary of sharing technical secrets with America, at least so long as it remained stingy in its support of Britain's war effort.

While Roosevelt's note was short on details such as who would possess bombs and how they might be used, his offer represented an opportunity for the British to enter what would become the Manhattan Project on an almost equal footing with America. But Churchill let several weeks elapse before offering a perfunctory response that Anderson and Lindemann had been delegated to speak with an OSRD representative in London. That meeting took place on November 21, with the British representatives giving the distinct impression that they believed themselves to be in the dominant position and that American security might not be as tough as it should be; Anderson advised Churchill to give the President a general assurance of the British desire to collaborate. But just over two weeks later, at a critical OSRD meeting on December 6, it was decided to greatly expand the American project; the next day, Japan attacked Pearl Harbor.

Churchill and Roosevelt soon met for the First Washington Conference to discuss war strategy (Dec. 22, 1941–Jan. 14, 1942), but they do not appear to have discussed the bomb at that time. Admittedly, in the immediate pressure of the war, such a thing must have seemed a distant possibility at best. Soon after Churchill's departure, Roosevelt approved Vannevar Bush's proposal for a much-expanded and reorganized nuclear project. Wallace Akers toured American project sites in early 1942, and came to realize how far ahead American scientists were on the experimental side of things. Upon returning to Britain, he proposed to Anderson and Lindemann that the British project should be merged with the American one; Anderson wrote to Bush that he felt it desirable to continue collaboration. Bush responded with an evasive description of administrative rearrangements in the US, but made no commitments.

As the American program ramped up between the fall of 1941 and late 1942, the idea of a cooperative program with Britain cooled considerably. By the spring of 1942, Roosevelt was urging Bush to push the project "with due regard to time"; the American program began gaining momentum and would soon come under the command of the Army. Britain's traction was slipping.

In June 1942, Churchill traveled to the United States for the Second Washington Conference. He discussed the uranium issue with Roosevelt in a private meeting at Roosevelt's family estate in Hyde Park, New York, on the afternoon of June 20, urging that Britain and America should pool their information, work as equal partners, and share whatever results might emerge, despite the fact that production plants would be located in the United States. Three weeks later, Roosevelt informed Bush that he and the Prime Minister were "in complete accord," but no written agreement had been signed nor any details specified. Roosevelt was a master political tactician; his words had no force of law. By this time Wallace Akers had managed to convince Lindemann of his point of view, and on July 30 Anderson wrote a pleading memo to Churchill in which he advocated merging the projects to capitalize on what assets the British could still contribute. With the Hyde Park "understanding" likely in mind, Churchill agreed. On August 5, Anderson attempted to formalize the discussions in letters to Bush, suggesting that a British-designed diffusion plant be built in America, that a heavy-water pile program be transferred to Canada, and that a joint nuclear energy commission be established. But by this juncture the initiative had been lost: The American program was in the middle of its transfer to military authority and its attendant secrecy; for General Groves and Vannevar Bush, international negotiations could only be a complication. While the British had made progress with diffusion, research on all other enrichment methods and reactor development were strictly American affairs. On October 1, Bush informed Anderson of the evolving arrangements in America, loosely referring to keeping up contact on how best to put the resources of both countries to work.

The diverging viewpoints of British and American atomic-project leaders became clear during late 1942 and early 1943 when Wallace Akers traveled to America to confer with James Conant. During a meeting on December 11, Conant presented the American perspective, which was that interchange should be restricted only to information that Britain could use during the war. Akers argued that Roosevelt and Churchill intended collaboration in both research and production, and felt that British scientists should have access to all large-scale American developments. Conant reported back to Bush the next day. However, four days thereafter, Bush carried to Roosevelt a 29-page document that had been prepared by a Military Policy Committee that had been established in September when the Army took over the project; the report recommended no or only limited interchange with the British. Roosevelt set the policy at limited interchange: Cooperation in the design and construction of the diffusion plant, research-level information on plutonium and heavy water, and no sharing of information on other methods of enrichment or Los Alamos.

Akers was bluntly informed of the new policy in a meeting on January 13, 1943. Churchill brought up the issue with Roosevelt again when the two met at the Casablanca Conference (January 14–24, 1943). Roosevelt left it to Bush to develop a reply, which was that there was no reason to change the American position. Churchill raised the issue yet again during a visit to Washington in late May, during which Bush was brought into the discussions with presidential aide Harry Hopkins and British advisors. On the rationale that since a weapon might be developed in time for use in the war (in which case the "direct use" scenario alluded to above would hold), Churchill departed with the understanding that he had secured a promise from Roosevelt that the work was to be joint and that interchange would be resumed. Bush met with Roosevelt on June 24 to review the situation. However, Roosevelt did not speak of his promise to Churchill, and Bush left the meeting with the impression that Roosevelt had no intention of going beyond the standing limited-interchange policy. Churchill raised the issue with Roosevelt yet again on July 9, at which time the President finally acquiesced: On the 20th he wrote to Bush to instruct him to renew full interchange with the British.

At the time of Roosevelt's July 20 directive, Bush was in London conferring with counterparts there on scientific aspects of the war. Roosevelt's note had not arrived when Bush met with Churchill on the 15th; the Prime Minister was furious that interchange seemed to have stalled. Unaware of the President's directive, Churchill, Hopkins, Bush, and Anderson met again on July 22, at which time Churchill offered a five-point proposition that would form the basis of the so-called Quebec Agreement that would be signed a month later. The essential points were that (i) the enterprise would be joint with free interchange; (ii) neither government would employ nuclear weapons against the other; (iii) neither would pass information to other countries without the consent of the other; (iv) use of the bomb in war would require common consent; and (v) the President might limit commercial or industrial uses by Britain in such a manner as he considered fair in view of the expense being borne by the United States. In Washington in early August, Bush and Conant met with Anderson again at the British Embassy to finalize the text of the agreement. Points (ii)–(v) would go into the Quebec Agreement essentially unchanged, but the interchange issue was still sticky. As a compromise, Anderson suggested the establishment of a "Combined Policy Committee" of American-British-Canadian representatives to coordinate what work would be done in each country and to serve as a focal point for exchanging information. Interchange on scientific research and development was to be "full and effective," but interchange in the area of design, construction and operation of full-scale plants was left on an ad-hoc basis to be decided by the Committee. In effect, the Committee was a bureaucratic dodge that let the Americans decide what information they would release; functionally, the committee did not set policy and met only eight times over two years.

The formal agreement was signed by Roosevelt and Churchill on August 19, 1943, during a meeting in Quebec City. As part of the interchange program, groups of British scientists, both native and newly-naturalized, went to America and Canada: the "British Scientific Mission" described previously.

The British contingent at Los Alamos was very highly-regarded. Hans Bethe was of the opinion that [quoted in Fakley (1983)]

> For the work of the Theoretical Division of the Los Alamos Project during the war the collaboration of the British Mission was absolutely essential. . . It is very difficult to say what would have happened under different conditions. However, at least, the work of the Theoretical Division would have been very much more difficult and very much less effective without the members of the British Mission, and it is not unlikely that our final weapon would have been considerably less efficient in this case.

General Groves' dismissive attitude toward British and Canadian contributions to the Manhattan Project was surely driven by patriotic pride, but was unfair; indeed, he was not above making use of scientists from those and other countries when he felt that they could contribute to the work. Unfortunately, the contributions of the British Mission to the success of the Manhattan Project are largely unappreciated. Politics can be much more complicated than physics.

References

Chadwick, J. (1969). Interview with Charles Weiner, April 20, 1969. Transcript available at https://www.aip.org/history-programs/niels-bohr-library/oral-histories/3974-4.

Chadwick, T., & Chadwick, M. (2021). Who invented the Trinity nuclear test's Christy gadget? Patents and evidence from the archives. *Nuclear Technology, 207*(S1), S356–S373.

Clark, R. W. (1961). *The birth of the bomb: The untold story of Britain's part in the weapon that changed the world*. London: Phoenix House.

References

Clark, R. W. (1965). *Tizard*. London: Methuen.

Close, F. (2020). *TRINITY: The treachery and pursuit of the most dangerous spy in history*. Penguin Books.

Dalitz, R. H. (2001). Marcus Laurence Elwin 'Mark' Oliphant. *Physics Today, 54*(7), 73–74.

Fakley, D. C. (1983). The British mission. *Los Alamos Science, 4*(7), 186–189.

Farmelo, G. (2013). *Churchill's bomb. How the United States overtook Britain in the first nuclear arms race*. New York: Basic Books.

Frisch, O. R. (1979). *What little I remember*. Cambridge: Cambridge University Press.

Gowing, M. (1964). *Britain and atomic energy 1939–1945*. London: St. Martin's Press.

Groves, L. (1983). *Now it can be told: The story of the Manhattan Project*. Cambridge, MA: Da Capo Press.

Lee, S. (2022). "Crucial? Helpful? Practically nil?" Reality and perception of Britain's contribution to the development of nuclear weapons during the second world war. *Diplomacy and Statecraft, 33*(1), 19–40.

Malenfant, R. E. (2005). *Experiments with the dragon machine*. Los Alamos Publication LA-14241-H. http://www.osti.gov/energycitations/purl.cover.jsp?purl=/876514-I1Txj9/

Moore, R. (2021). Rudolf Peierls' "Outline of the development of the British tube alloy project": A 1945 account of the earliest UK work on atomic energy. *Nuclear Technology, 207*(S1), S374–S379.

Peierls, R. (1985). *Bird of passage: Recollections of a physicist*. Princeton: Princeton University Press.

Ruane, K. (2016). *Churchill and the bomb in war and cold war*. London: Bloomsbury.

Szasz, F. M. (1992). *British scientists and the Manhattan Project: The Los Alamos years*. New York: St. Martin's Press.

Chapter 3
The Memorandum: Qualitative Part

Abstract This brief chapter describes the qualitative part of the Frisch-Peierls memorandum. While occupying only three pages, this document touched on every aspect of nuclear weapons: Their likely destructive and radiological effects, how they could be constructed and triggered, the need to industrialize isotope enrichment, the military and ethical issues that would accompany such weapons, and what might be happening in Germany.

This relatively brief chapter offers commentary on the qualitative part of the Frisch-Peierls memorandum, which is reproduced in Appendix A. The discussion is grouped by paragraphs in the memorandum that concern a common theme.

Paragraphs [1]–[3]: Destructive Effects of a Nuclear Weapon

Frisch and Peierls (hereafter F & P on occasion) get right to the point regarding the likely power of a bomb based on utilizing "the energy stored in atomic nuclei." They estimate the energy released by such a "super bomb" to be equivalent to the explosion of 1,000 tons of dynamite, what we would now term a kiloton. This would prove in time to be a serious underestimate; the Hiroshima and Nagasaki bombs are estimated to have been equivalent to about 13 and 25 kilotons, respectively; these are now considered to be small yields. In World War II, a common bomb weight was 500 pounds, but this was the total weight of the weapon including its metal case and firing mechanism; the actual amount of explosive was about 200 pounds. To a military planner of that era, a bomb that would be equivalent to 1,000 tons of explosive would have seemed like science fiction. Could a mission requiring hundreds of planes, each carrying a dozen or two bombs, really be replaced by one bomber carrying one weapon?

F & P estimated that the resulting explosion would probably cover the center of a big city. At Hiroshima, almost everything within a mile of ground zero would be destroyed except for the most heavily reinforced buildings; the firestorm created by the bomb burnt out about 4.4 square miles, and fires were ignited up to 15,000 feet away. At Nagasaki, the more powerful plutonium-based bomb was used, but the local hilly geography acted to contain the destructive effects; nevertheless, an area of about three square miles was totally destroyed, and charring of telephone poles was observed to a distance of 11,000 feet. F & P did not exaggerate.

As F & P point out, what makes a nuclear weapon qualitatively different from an ordinary explosive is the immense amount of radioactive fallout created by direct fission products and the remnants of the bomb casing being rendered radioactive by capturing neutrons and thereby inducing beta-decays. Fallout can be spread by wind and rain; exposure patterns can be quite variable. At Hiroshima and Nagasaki, most of the estimated total of 100,000 people killed succumbed to the blast, fire, and heat effects of the bombs; estimates of the fraction of radiation-caused deaths run to about 15%. Of course, many victims likely suffered multiple effects.

Paragraph [4]: Need for ^{235}U and Isotope Separation

F & P drastically underestimated the effort that would be necessary to isolate the necessary amount of fissile uranium-235 needed to make a bomb; they suggested that a few cwt of natural uranium could serve as the raw material. A "cutweight" (cwt) is a somewhat archaic unit of weight used in commodity trading equivalent to 100 pounds. Their estimate would seem reasonable on the basis that they estimated the critical mass to be about a pound and that ^{235}U comprises about 0.7% of natural uranium. However, isotope separation would prove to be a very difficult and inefficient business; the wartime Manhattan Engineer District would source several thousand tons of uranium ores from Africa, Canada, the United States and Germany (captured materials) to feed enormous isotope separation plants in Oak Ridge, Tennessee and plutonium-synthesizing reactors in Hanford, Washington. An aspect of this that was unanticipated when F & P wrote their memorandum was that one does not build such factories to build just one or two bombs; by their nature, they become mass-production enterprises. Later in the memorandum (paragraph [10]), Frisch and Peierls suggest that a method of isotope separation invented by Professor Klaus Clusius of the University of Munich could be used. Clusius' method, known as liquid thermal diffusion, would be used during the Manhattan Project, but on a scale much smaller than other methods. Ultimately, plant and operations costs for isotope enrichment at Oak Ridge would amount to some $1.2 billion, over half of the total cost of the project. Acquisition of uranium would consume over $20 million alone.

Paragraphs [5] and [6]: Critical Mass and Triggering

These two paragraphs discuss the technical issues of critical mass and how to trigger a nuclear weapon. F & P estimated the critical mass as about a pound (a tremendous underestimate, as explained in Chap. 4), but point out that since an amount of material less than the critical mass will be quite harmless, this suggests a method for transporting a bomb to a target: Simply divide the fissile material into two pieces that are each smaller than one critical mass, keep them apart by a few inches, and arrange to bring them together when the bomb is to be triggered. To initiate the chain reaction, they planned to rely on the fact that the constant flux of cosmic rays impinging on the Earth from outer space contains high-energy neutrons; one would likely have to wait for but only a second or so for one of these to do the job. Ultimately, neutron-emitting "initiator" devices were developed for a more controlled approach, but the cosmic ray method would be a quite sensible one for a crude terrorist bomb.

This idea of two separated pieces of fissile material is exactly what was used in the uranium-based Hiroshima "Little Boy" bomb as sketched in Fig. 3.1. Ordnance engineers mounted the barrel of a naval artillery cannon inside a bomb casing, and placed one piece of uranium, the "target piece," a cylinder made up of rings of material, at the nose end of the barrel. The second piece, the "projectile piece," was a cylindrical sleeve of rings placed at the tail end. When radars and barometers mounted on the bomb indicated that it had fallen to a pre-programmed detonation height, a conventional charge was set off to propel the projectile piece into the target piece; together, the two comprised more than a critical mass. In the figure, the tamper serves to reflect escaping neutrons back into the assembled core to enhance the efficiency of the explosion; F & P did not consider a tamper in their analysis.

Paragraph [6] touches on an issue that would become very important at Los Alamos: That the assembly of the two pieces must happen quickly. This is because, at some point during the assembly, a critical mass will come to be present. If a stray neutron should initiate the chain reaction before assembly is complete, the timescale for the reaction is so short before it shuts itself down due to expansion of the core—on the order of a microsecond—that the resulting explosion might be very inefficient, although likely enough to destroy the bomb and so deny an enemy the fissile material. Since the initiating cosmic rays cannot be controlled, there is always some probability of this happening. F & P estimated, without any explanation, that the probability of a bomb failing in this way would be about one out of 100. Estimating this probability is not a trivial issue; a very approximate calculation appears in Sect. 4.8.

Two other issues along this line are relevant. Look back to Sect. 1.1 where (α, n) reactions are described: An alpha-particle emitted by a radioactive nucleus strikes the nucleus of a light element, liberating a neutron. If your fissile material is contaminated with light elements, say due to some chemical processing during the isotope separation, this is another possible source of a premature detonation. You want your fissile material to be as pure as possible; Frisch and Peierls remark on this specifically in the technical part of the memorandum. Similarly—and in addition—fissile elements do spontaneously fission. This was not known to F & P at the time they wrote the memorandum. As described in Chap. 2, Frisch later observed spontaneous fission of uranium; discovery of this effect is usually attributed to Georgy Flerov

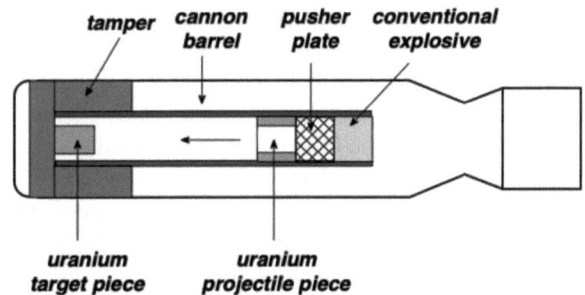

Fig. 3.1 Schematic illustration of a gun-type "Little Boy" fission weapon like that used at Hiroshima. Sketch by author

and Konstantin Petrjak (sometimes spelt Petrzhak) in Russia in June 1940; they published their result in the American journal *Physical Review* (Flerov and Petrjak 1940). Spontaneous fission of plutonium would prove such a serious issue that an entirely different triggering mechanism would be required, but this is beyond the scope of this book. The bottom line is that when assembling a supercritical core, one wants to do it as fast as possible.

Paragraphs [7]–[9]: Military and Ethical Considerations

The seventh paragraph of the memorandum is one short sentence in which F &P remark that they did not feel competent to discuss the strategic value of such a bomb. In paragraphs eight and nine, however, they do exactly that in direct, lucid prose utterly free of the jargon such a document would likely contain today. They describe such a bomb as "practically irresistible" given that nothing could be expected to withstand such an explosion. There are caveats, however: Radiation would prevent anyone—defender or attacker alike—from entering the affected area for some days. In paragraph nine they soberly point out the radiation would likely kill large numbers of civilians, "and this may make it unsuitable as a weapon for use by this country." However, since at least the German bombing of Coventry in November 1940, both sides were systematically targeting civilians, if not admitting doing so. Interestingly, they suggest using the bomb as a depth charge near a naval base, although this too would likely cause extensive civilian casualties by flooding and radiation. The idea of using such a bomb against an enemy port was also raised in the August 1939 letter to President Roosevelt signed by Albert Einstein.

Paragraphs [10]–[12]: Germany and Deterrence

These three paragraphs turn to the issue of what might be happening in Germany. It is remarked that since all of the theoretical data (and, in retrospect, a growing body of experimental data) bearing on the possibility of a nuclear weapon were openly published, it was entirely possible that such a weapon could be being developed in Germany. Indeed, in some ways the Germans were at that point ahead of the allies. From the spring of 1939, there were briefly two research programs underway in Germany; these were merged into one known as the "Uranium Club" under the Army's Ordnance department in September, and a ban was placed on exports of uranium compounds. Rich uranium ores were known to exist in Czechoslovakia, and F & P advised that it would be helpful to know what was happening with mines there; the Einstein-Roosevelt letter also mentioned that Germany had halted the sale of uranium from Czech mines. F & P also remark that it would be helpful to know of Professor Clusius' whereabouts, as he would likely be involved in any isotope separation efforts. Ultimately, the German program would be hobbled by personality conflicts, disputes over resources, lack of interest from top government levels, and the effects of allied bombing raids, but F & P knew from personal experience that Germany had many competent physicists and engineers and a very advanced chemical industry. It has been estimated that the total funding for the German program amounted to about $2 million, or about 1/1000 of what was spent on the Manhattan Project. Surveys of the German nuclear program can be found in Mark Walker's *German National*

Socialism and the Quest for Nuclear Power 1939–1949 (Walker 1989) and *Hitler's Atomic Bomb: History, Legend, and the Twin Legacies of Auschwitz and Hiroshima* (Walker 2024); see also Popp (2021) and Popp & de Klerk (2023) for analyses of some of the internal complications of the German program.

F & P also anticipated nuclear secrecy. Paragraph 11 points out that if nobody in Germany had thought of the possibility of a super bomb, it was extremely important to keep their report secret lest any rumours leak out.

Paragraph 12 is uncanny in its anticipation of the later Cold War, pointing out that if Germany was working on such a weapon, the only effective reply would be a counter-threat with a similar bomb. They urged that work begin promptly, "even if it is not intended to use the bomb as a means of attack." The concept of nuclear deterrence was born in Birmingham in March 1940. Given that the requisite isotope separation would likely require several months, they felt that "the matter seems, therefore, very urgent." At Oak Ridge, Tennessee, the better part of two years would be required to isolate enough ^{235}U for the first bomb.

In 1944 and 1945, American scientists would echo Frisch and Peierls in the so-called Jeffries and Franck reports, which were prepared to address how nuclear weapons should be introduced to the world (direct use or a demonstration), and how issues of postwar control of such weapons and fissile materials would need to be crafted to forestall a potentially catastrophic arms race or the possibility of terrorists acquiring such weapons, while also encouraging legitimate peaceful research. But by that time the military momentum behind the project and the state of the war made direct military use an essentially foregone conclusion.

Paragraphs [13]–[15]: Radiological Protection

The radioactivity that would be generated in a nuclear explosion was obviously of great concern to F &P. In these three paragraphs, they propose that specially-trained detection squads would have to be available to enter a contaminated area by either aircraft or shielded automobiles, equipped with suitable instruments and self-contained oxygen supplies. Their job would be to determine when and for how long people could enter the affected area, a matter which required further research; such enormous amounts of radioactive exposure would never be encountered in a conventional laboratory situation.

Paragraph [16]: Conclusion and Need for Further Research

In what must be one of the greatest understatements in history, F & P remark that their conclusions are not based on direct experiments, "... since nobody has ever yet built a super bomb..." However, the relevant nuclear physics was well-established. The main uncertainty lay in estimating the critical mass, an issue that could be clarified as soon as a small amount of uranium was separated and appropriate experiments undertaken. As described in Chap. 1, this is precisely what Alfred Nier and his collaborators were doing in America at virtually the same time.

In their brief document, Frisch and Peierls touched on physics, military strategy, the need for intelligence, the spectre of fallout, the ethics of nuclear warfare, the possibility that a nuclear weapon could uncontrollably pre-detonate, deterrence, and the need for further research to better pin down the possibilities. It is no wonder that in her 2003 essay, British historian Lorna Arnold remarked that in her over 30 years of researching nuclear history, she knew of no text so fascinating and impressive as the Frisch-Peierls memorandum.

References

Flerov, G. N., & Petrjak, K. A. (1940). Spontaneous fission of uranium. *Physical Review, 58*(1), 89.

Popp, M. (2021). Why Hitler did not have atomic bombs. *Journal of Nuclear Engineering, 2*(1), 9–27.

Popp, M., & de Klerk, P. (2023). The Peculiarities of the German Uranium Project (1939–1945). *Journal of Nuclear Engineering, 4*(3), 634–653.

Walker, M. (1989). *German national socialism and the quest for nuclear power 1929–1949.* Cambridge: Cambridge University Press.

Walker, M. (2024). Hitler's atomic bomb: History, legend, and the twin legacies of Auschwitz and Hiroshima. Cambridge: Cambridge University Press.

Chapter 4
The Memorandum: Technical Part

Abstract This chapter is the heart of this book: A paragraph-by-paragraph analysis of the physics behind the quantitative part of the Frisch–Peierls memorandum. Topics include the nature of uranium isotopes, slow and fast-neutron chain reactions and their timescales, why a bomb requires a fast-neutron chan reaction, critical mass, likely energy yield, the danger of weapon predetonation, how uranium enrichment could be realized, and radiation effects. Their document still makes for a compact primer on the physics of nuclear weapons.

This chapter dissects the technical part of the Frisch–Peierls memorandum, which is reproduced in Appendix B.

The 24 paragraphs of the technical appendix touch on practically every aspect of nuclear weapons: The response of the two uranium isotopes to neutron bombardment, why a bomb based on a moderated slow-neutron reaction would be impractical, the possibility of a fast-neutron bomb utilizing pure ^{235}U, the critical mass, the speed and yield of the chain reaction, how a bomb could be triggered, the possibility of predetonation, thermal diffusion as a means of separating isotopes, and the likely effects of radioactive fallout.

Two items to note: In this chapter, I revert to the traditional format for citing publications as there are many occasions where specific papers from which Frisch and Peierls might have adopted values for various parameters are referenced. Also, Frisch and Peierls primarily used cgs units as opposed to the current SI convention; for comparison with their values I also use cgs, but give SI equivalents in some occasions.

In what follows, I treat the paragraphs grouped according as their related themes in the order in which they occur in the memorandum. Section 4.5 on the critical radius and mass is divided into two subsections.

4.1 Paragraphs [1] and [2]: Uranium and Neutron Bombardment

The technical part of the memorandum opens with the surprising statement that the idea of a "super bomb" "has been discussed a great deal", although no names are named. Given that this is followed by arguments which seemed to exclude the possibility, F & P likely had Niels Bohr in mind. As described in Sect. 1.3, Bohr had touched on the issue of the potential fast-neutron fissility of ^{235}U, but had not emphasized the point. F & P then remark that they wish to point out and discuss a possibility that had theretofore been overlooked, which is precisely this idea.

Paragraph [2] summarizes what can happen when uranium nuclei are struck by neutrons: scattering, non-fission capture, and fission with an energy release of ~ 200 MeV. Their remark that the neutron loses energy if its energy is above about 0.1 MeV and scatters can be understood by looking back to Fig. 1.5. In natural-abundance uranium, the situation will be dominated by the huge preponderance of ^{238}U. At about 0.1 MeV, the capture and energy-robbing inelastic scattering cross sections for 238 are about equal; above this energy, the inelastic scattering cross section increases sharply while the capture cross section declines. Review Sect. 1.4 for a discussion of why this effect and the reaction timescale effect discussed in Sect. 4.4 below renders natural uranium impractical for a fission bomb.

4.2 Paragraphs [3]–[5]: Chain Reactions

This group of paragraphs brings us to the first quantifications of the technical appendix. Paragraph [3] remarks that since the number of neutrons emitted per fission is greater than one, a chain reaction should be possible. This quantity is traditionally given the symbol ν, and they quote a most probable value of $\nu = 2.3$ based on two independent measurements.

Soon after the discovery of fission, various researchers began looking for secondary neutrons, and proof of their existence was not long in emerging. In a paper published in the April 15, 1939 edition of *Physical Review*, Anderson, Fermi, and Hanstein estimated "about two" neutrons emitted per each one captured; (Anderson et al., 1939b). Essentially simultaneously and published in the same edition of the *Physical Review*, Leo Szilard and Walter Zinn came to the same conclusion (Szilard & Zinn, 1939). By July, Anderson, Fermi, and Szilard had arrived at a refined value of 2.2 (Anderson et al., 1939c), and in August a separate experiment by Zinn and Szilard (1939) yielded an estimate of 2.3. There do not seem to have been any measurements of ν in Britain by this time. These figures are in very respectable agreement with the present-day value of $\nu = 2.57$; see Chadwick (2021) p. S32.

Paragraph [4] reiterates how the inelastic scattering/capture behavior of ^{238}U prohibits setting up a chain reaction in ordinary uranium.

4.2 Paragraphs [3]–[5]: Chain Reactions 59

Paragraph [5] makes a brief remark that various researchers had attempted to construct chain reactions using ordinary water as a moderator. These efforts occurred in five countries in the months following the discovery of fission up to about the time of the memorandum: France, Britain, Canada, Germany, and the United States. Here I briefly describe those that appeared in the open literature, or, in the case of Britain, of which Frisch and Peierls might have been aware. An extensive inter-country comparison of wartime pile development can be found in Reed (2021b).

In Paris, Hans von Halban and collaborators undertook the first pile experiments essentially simultaneously with those of Herbert Anderson and Enrico Fermi at Columbia University in New York. Between March 1939 and January 1940, the Paris group constructed six piles. These involved liquid uranium compounds of up to $\sim$300 kg of uranium oxide, plus water or cubes of paraffin as a moderating agent, all contained within copper spheres which could be immersed in water. They detected net neutron generation with a sphere containing a solution of uranyl nitrate, but no self-sustaining reaction; this was published in the March 18, 1939 edition of Nature. In late 1939 they constructed a Fermi-like graphite pile, five feet square by ten feet high, to determine the neutron capture cross-section of graphite. Results were inconclusive, but hinted that a heterogeneous lattice-type arrangement of uranium and moderator as would eventually be used by Fermi might be feasible. The last Paris experiment (January 1940) involved a lattice-type arrangement of dry oxide plus paraffin cubes as a comparison against earlier wet-oxide homogeneous arrangements; a decrease in neutron capture with the heterogeneous configuration was noted, as would be verified by Fermi. The overall conclusion of these experiments, however, was that water captured too many neutrons to serve as a moderator. References are Halban et al. (1939a, b); an excellent summary of the French work can be found in Spencer Weart's *Scientists in Power* (Weart, 1979).

In London, in the summer of 1939 George P. Thomson of the MAUD committee undertook experiments with about a ton of uranium oxide contained within a cast-iron sphere, trying both water and paraffin as moderators; he concluded that no self-sustaining chain reaction would be possible with such moderators. Thomson does not seem to have published this result; as Britain was near to war, this is not at all surprising, but it can be speculated that Frisch and Peierls might have heard of the work. Thomson's work is described in Ronald Clark and Margaret Gowing's books on the British nuclear program; Clark (1961) and Gowing (1964). Clark also describes some of the French efforts.

The American nuclear pile program is of course well-documented. Before wartime censorship went into effect, some papers were published in the *Physical Review*, but the essential source of information here is Volume II of Enrico Fermi's collected works; Fermi (1965). This volume covers the time from Fermi's arrival in the United States in early 1939 to his death in 1954, and includes copies of reports to various administrative bodies such as the Manhattan Project's Uranium Committee and the National Defense Research Committee, which in mid-1940 took on responsibility for the project.

Pile experiments under Fermi's supervision began soon after his arrival at Columbia University in early 1939. His first efforts were in collaboration with Herbert Anderson, whom we have already met, and H. B. Hanstein. Their first experiment involved a cylindrical tank which would have held $\sim$570 kg of water if completely filled and into which a neutron source could be inserted. A glass bulb could be placed in the tank; when it was filled with uranium oxide, an increase in the neutron population above that when no oxide was present was detected: Fissions were occurring. This is the Anderson, Fermi, and Hanstein paper cited above wherein they estimated that each fission generated about two secondary neutrons.

The second Columbia experiment resulted in the Anderson, Fermi, and Szilard paper also cited above. This used cylindrical cans of dimensions 5 cm diameter by 60 cm high which contained a total of $\sim$200 kg of U_3O_8 immersed in $\sim$540 liters of a manganese sulfate solution; this was presumably all contained within the same cylinder as the preceding experiment. The manganese served as a neutron detector via its propensity to capture neutrons and become radioactive. Their conclusion was that a chain reaction "could be maintained in a system in which neutrons are slowed down without much absorption until they reach thermal energies ...", but with the caveat that "it remains an open question, however, whether this holds for a system in which hydrogen [i.e., water] is used for slowing down the neutrons." Overall, however, they were skeptical that even an optimal mixture of water and uranium would be able to sustain a chain reaction. Fermi and his group would conduct no further water-based experiments; they moved to graphite as a moderator. Further piles were built at Columbia from September 1940 onwards until Fermi moved to Chicago in early 1942, but these were described in classified reports that would not have been available to Frisch & Peierls.

By the time Frisch and Peierls were preparing their memorandum, openly-published results were strongly suggesting that no slow-neutron chain reaction with ordinary uranium and water would work; as described in Sect. 1.4, a water-moderated reactor requires slightly enriched uranium.

4.3 Paragraph [6]: No Slow-Neutron Bomb

Paragraph [6] quantifies the issue of the uselessness of a slow-neutron reaction for a bomb by examining the timescale involved. Reaction timescales are an important part of the overall discussion here; they will in particular reappear in Sect. 4.6 below. Appendices C and D also contain much relevant material.

To begin, it is helpful to establish some key concepts and convenient formulae.

The timescale of a reaction is dictated by two quantities: The mean free path for the reaction and the neutron speed involved.

The mean free path is the average distance that a neutron will travel before interacting with a target nucleus for a particular reaction being considered (scattering, fission, non-fission capture ...). This depends on the density of the material involved

4.3 Paragraph [6]: No Slow-Neutron Bomb

and the cross section for the desired reaction. The mean free path is always designated by the symbol λ. A derivation can be found in any nuclear physics text; the expression is

$$\lambda = \frac{1}{\sigma n}. \tag{4.1}$$

In this expression, σ is the cross section involved and n the number density of nuclei, that is, the number of fissile nuclei per unit volume. If the density of the material is ρ gr cm^{-3} and it has atomic weight A gr mol^{-1}, this is given in nuclei per cubic centimeter by

$$n = \frac{\rho N_A}{A}, \tag{4.2}$$

where N_A is Avogadro's number, 6.022×10^{23} objects (here, nuclei) per mole.

For the materials we will be dealing with, number densities tend to be on the order of 10^{22} per cubic centimeter. To this end, define n_{22} as the nuclear density expressed in such units. Now, cross sections are conventionally measured in barns, with 1 bn $= 10^{-24}$ cm^{-2}. Use σ_{bn} to represent a cross section in barns. We can consequently cast Eq. (4.1) into a form that gives λ directly in centimeters; call this λ_{cm}:

$$\lambda_{cm} = \frac{100}{\sigma_{bn} \, n_{22}}. \tag{4.3}$$

Neutrons will always have a distribution of speeds; call the average speed in centimeters per second $\langle v \rangle$; see Eq. (4.5) below. If a neutron has to travel distance λ_{cm} at this speed, ordinary kinematics tells us that the time taken will be $\lambda_{cm}/\langle v \rangle$. This timescale is designated as τ_o:

$$\tau_o = \frac{\lambda_{cm}}{\langle v \rangle} = \frac{100}{\langle v \rangle \sigma_{bn} \, n_{22}}. \tag{4.4}$$

For a convenient expression for $\langle v \rangle$, neutrons are usually specified by their $mv^2/2$ kinetic energy in MeV. Looking up the mass of a neutron and recalling that 1 MeV $= 1.6022 \times 10^{-13}$ Joules, you should be able to show that the speed in cm per second is

$$\langle v \rangle = 1.383 \times 10^9 \sqrt{E_{MeV}}. \tag{4.5}$$

Frisch and Peierls quote two slow-neutron timescales. The first one is a slowing-down time of 10^{-5} s, and the second a subsequent time-to-fission of 10^{-4} s. Neutrons emitted in fissions have an average energy of 2 MeV (this is the modern value; they used ~0.5 MeV). Neutrons get slowed by successive collisions with moderator nuclei; here we concern ourselves with a water moderator, as did F & P.

At a nuclear level, scattering is a fundamentally quantum process, but a classical model will serve us perfectly well for establishing an order-of-magnitude estimate. The analysis presented here is adapted from Reed (2020b), which presents an undergraduate-level analysis of neutron slowing by successive elastic collisions with nuclei. In this analysis, a neutron of one mass unit successively strikes and scatters from nuclei of mass A mass units. For the case of water as a moderator, we concern ourselves with hydrogen nuclei, which contribute the dominant slowing effect. The struck nuclei are assumed to be initially stationary. It is shown in this analysis that for a single collision, the ratio of the neutrons' kinetic energy after the collision to that before is given by

$$\frac{K_{post}}{K_{pre}} = 1 - \left[\frac{4A}{(A+1)^2}\right]\cos^2\psi,$$

where ψ is the angle between the neutron's original direction of motion and the direction of the force exerted on it by the struck nucleus; $\pi/2 \leq \psi \leq \pi$. This will vary from collision to collision, but for sake of simplicity I assume that each collision is characterized by the average value of $\cos^2\psi$ over this range, which is exactly 1/2. For one collision with hydrogen ($A = 1$) we then get

$$\frac{K_{post}}{K_{pre}} = 1 - \left[\frac{4}{2^2}\right]\frac{1}{2} = \frac{1}{2},$$

that is, the neutron will lose about one-half of its kinetic energy with each collision.

Now suppose that a neutron suffers N_c successive such collisions, taking its kinetic energy from $K_{initial}$ to K_{final}. How many collisions will this require? We must have

$$\frac{K_{final}}{K_{initial}} = \left(\frac{1}{2}\right)^{N_c},$$

that is

$$N_c = \frac{\log(K_{final}/K_{initial})}{\log(1/2)}.$$

Apply this to neutrons that begin with $K_{initial} = 1$ MeV and have to be thermalized to $K_{final} = 0.025$ eV:

$$N_c = \frac{\log(0.025/1 \times 10^6)}{\log(1/2)} \sim 25.$$

Now, since $K_{post}/K_{pre} = 1/2$, the ratio of the post- to pre-collision speed after any one collision will be $v_{post}/v_{pre} = 1/\sqrt{2}$. About how long will the N_c collisions require? Imagine that a neutron beings with speed v_o. After one collision its speed will be $v_1 = v_o/\sqrt{2}$. After two collisions its speed will be $v_2 = v_1/\sqrt{2} = v_o/(\sqrt{2})^2$, and so on. Now say that the mean free path that has to be covered between each collision is λ; there will be a comment on this below. The time required from the fission-birth

4.3 Paragraph [6]: No Slow-Neutron Bomb

of the neutron to the first collision will be λ/v_o, that from the first collision to the second collision will be λ/v_1, and so forth. Adding up the times for all N_c collisions gives

$$t_{slow} = \frac{\lambda}{v_o} + \frac{\lambda}{v_1} + \frac{\lambda}{v_2} + \cdots + \frac{\lambda}{v_{N_c}}$$

$$= \lambda \left[\frac{1}{v_o} + \frac{(\sqrt{2})}{v_o} + \frac{(\sqrt{2})^2}{v_o} + \cdots + \frac{(\sqrt{2})^{N_c}}{v_o} \right]$$

$$= \frac{\lambda}{v_o} \left[1 + (\sqrt{2}) + (\sqrt{2})^2 + \cdots + (\sqrt{2})^{N_c} \right].$$

The sum in square brackets here is a geometric series; the result is

$$t_{slow} = \frac{\lambda}{v_o} \left[\frac{2^{(N_c+1)2} - 1}{\sqrt{2} - 1} \right].$$

For our ~ 25 collisions this reduces to

$$t_{slow} \sim 20000 \frac{\lambda}{v_o}.$$

Now, an approximation that has been made here is that the mean free path is the same for each collision segment. This is strictly not true as the scattering cross-section is a function of neutron speed. However, this turns out to be not too bad an approximation in the case of hydrogen. At 1 MeV, the cross-section is about 4 barns; this rises to about 20 barns at 0.1 MeV, and remains essentially constant down to thermal energies. I will adopt 20 barns to apply over the entire slowing process; the vast majority of collisions will be at the lower end of the energy range.

The density of water is 1 gr cm^{-3}, and its atomic weight is 18 gr mol^{-1}. Equation (4.2) gives $n = 3.346 \times 10^{22}$ cm^{-3}, or $n_{22} = 3.346$. Doubling the scattering cross-section to 40 bn to account for two hydrogen atoms per water molecule gives, from Eq. (4.3) and rounding off slightly,

$$\lambda_{cm} = \frac{100}{\sigma_{bn} n_{22}} = \frac{100}{(40)(3.346)} \sim 0.75.$$

From Eq. (4.5), a 1-MeV neutron has a speed of $\sim 1.383 \times 10^9$ cm/s. The slowing time is then

$$t_{slow} \sim 20000 \left(\frac{0.75 \text{ cm}}{1.383 \times 10^9 \text{ cm/s}} \right) \sim 1.09 \times 10^{-5} \text{ s}.$$

Despite the various approximations that have been made, this is in remarkable agreement with Frisch and Peierls' estimate. One might choose a different starting energy for the neutron or a somewhat different cross-section, but the important thing is that the order-of-magnitude will evaluate to $\sim 10^{-5}$ s, as they claimed. This is a brief time, but as explained in the discussion of paragraph [11] below (Sect. 4.6), the time required for a *fast* neutron within uranium to cause a fission is on the order of a few times 10^{-9} s, four orders of magnitude shorter!

At this point, our neutrons have been slowed down, and now face the second time interval to cause a fission.

The origin of F & P's estimate of 10^{-4} s for a thermalized (not fast) neutron to cause a fission can be speculated as follows. A slowed neutron has entered a lump of *natural* uranium and is looking for a nucleus to fission. F & P took the density of uranium to be 15 gr cm^{-3}. The vast majority of nuclei will have $A = 238$, for which Eq. (4.2) gives $n_{22} = 3.795$. For $\tau_o = 10^{-4}$ and $\langle v \rangle = 2.2 \times 10^5$, Eq. (4.4) then demands $\sigma_{bn} \sim 1.2$. Anderson et al. (1939a) had estimated the fission cross-section for natural uranium with thermal neutrons as 2 bn and that for fast neutrons as about 0.1 bn, so using an estimate on the order of 1 bn is reasonable. Averaged across the spectrum of energies of fission-liberated neutrons, the present-day figures for these two values are about 4.2 and 0.32 bn, respectively; see Table 4.1. There will be much more on cross-sections in Sect. 4.5.2.

Leaping ahead here somewhat, we can use these present-day values to emphasize the point F & P are getting at, which is the tremendous contrast between timescales for slow and fast neutrons. Take a fast neutron to have $E_{MeV} = 2$. For neutrons traveling within material of the same density, Eqs. (4.4) and (4.5) show that the ratio of timescales will be

$$\frac{\tau_{slow}}{\tau_{fast}} = \frac{\sigma_{fast}\sqrt{E_{fast}}}{\sigma_{slow}\sqrt{E_{slow}}} \sim \frac{0.32\sqrt{2}}{4.2\sqrt{2.5 \times 10^{-8}}} \sim 680. \qquad (4.6)$$

The catch is that the analysis of criticality in Appendix D shows that the energy yield one can expect from a bomb before it blows itself apart is inversely proportional to the *square* of τ. This means that a slow-neutron bomb would release on the order of $(1/680)^2 \sim 2 \times 10^{-6}$ times the energy of a fast-neutron one equipped with the same amount of fissile material. If one of the latter is designed for a yield of 10 kt TNT equivalent (about that of the Hiroshima uranium bomb), the slow-neutron counterpart would have a yield of only ~ 40 pounds TNT equivalent, much less than even a small ordinary bomb. In effect, a slow-neutron bomb would mean attempting to drop a reactor on an adversary, with the resulting "explosion" being more of a fizzle.

Table 4.1 Modern fission cross section values (bn), Data from https://atom.kaeri.re.kr/old/ton/

Speed	^{235}U (bn)	^{238}U (bn)	Weighted by natural abundance
Fast	1.235	0.308	0.315
Thermal	584	–	4.208

4.4 Paragraphs [7]–[9]: Possibility of a Fast-Neutron Bomb

Having set aside the idea of a slow-neutron weapon, F&P turn to the possibility of a fast-neutron one. This is laid out in three paragraphs which summarize Niels Bohr's conclusions regarding slow and fast neutrons and the behaviors of the two isotopes ^{235}U and ^{238}U that are presented in Sects. 1.3–1.4 of the present volume. However, with their mix of comments on slow and fast neutrons, both isotopes, and then abandoning consideration of ^{238}U to focus on a fast-neutron reaction with ^{235}U, these passages would be confusing to a reader unfamiliar with this background.

Frisch and Peierls' essential conclusion here is that "This permits, in principle, the use of nearly pure ^{235}U in such a bomb, a possibility which apparently has not so far been seriously considered." Paragraph [9] emphasizes that ^{235}U will fission under bombardment by neutrons of any energy and that it will not be necessary to treat it with any moderating material. Indeed, from the timescale argument immediately above, F & P point out that a fast neutron reaction "... develops with very great rapidity so that a considerable part of the total energy is liberated before the reaction gets stopped on account of the expansion of the material." With this, the concept of a fast neutron fission weapon is born.

Paragraph [9] contains a curious but significant error. F & P state that "... from rather simple theoretical arguments it can be concluded that almost every collision produces fission ... ". This has the simplifying consequence that they can neglect scattering in their calculations, but in reality this is not so. As described by Pearson (2024), it is an inevitable consequence of the wave nature of particles that any non-elastic reaction process - for example, fission - must also lead to elastic scattering, and that the cross section for elastic scattering must equal to the total cross section for all non-elastic processes if the latter is maximal, as F & P assumed. Peierls especially would have been familiar with the relevant quantum theory, so one has to wonder what they might have been thinking. In March 1940, no inelastic scattering measurements were yet available, so there would have been no conflict with contemporary experimental evidence in assuming zero *inelastic* scattering cross section, but it does not follow that elastic scattering could also be neglected.

In a sense, neglecting scattering can be considered to be erring in the side of caution: Scattering acts to *decrease* the critical mass of a fissile core. This is because if a neutron that is otherwise destined to escape the core and be lost, scattering can redirect it to a path that would pass through more core material than if it had escaped directly, increasing its chance of causing a fission along the way. For fast neutrons, the modern data for ^{235}U indicate non-zero cross sections for all of fission ($\sim$1.24 bn), elastic scattering ($\sim$4.57 bn), and inelastic scattering ($\sim$1.80 bn). However, ^{235}U has no fission threshold, so inelastic scattering does not act to suppress a chain reaction. The modern value for the critical mass of ^{235}U is about 46 kg; if there were no scattering, this would be more than doubled.

4.5 Paragraph [10]: The Critical Radius and Mass

This is a key paragraph where the critical radius and mass are estimated, and where the overestimate of the fission cross section which has attracted so much attention occurs. Section 4.5.1 looks at the numbers for the critical radius and mass, and Sect. 4.5.2 possible reasons for the adopted cross section.

4.5.1 The Critical Radius

Frisch and Peierls state in their paragraph [10] that if scattering is disregarded and one adopts $\nu = 2.3$ neutrons per fission, then the critical radius is approximately 0.8 times the mean free path for fission, λ_f. A derivation of this appears in Appendix C; here we look at the resulting numbers.

Incorporating the factor of 0.8, Eq. (4.3) gives the critical radius R_{co} (see the notation of Appendices C and D) as

$$R_{co} = \frac{80}{\sigma_{bn}\, n_{22}}. \tag{4.7}$$

A reminder about units: R_{co} will be in centimeters when σ_{bn} is in barns and n_{22} is the multiple of 10^{22} nuclei per cubic centimeter; see Sect. 4.3. Frisch and Peierls adopted a density for uranium of 15 gr cm^{-3}; with $A = 235$, this density gives $n_{22} = 3.844$. They then adopt $\sigma_{bn} = 10$, and get a critical radius of less than an inch:

$$R_{co} = \frac{80}{(10)\,(3.844)} = 2.08\,\text{cm},$$

which they rounded to 2.1 cm. With their adopted density, this corresponds to 565 grams, or about one-and-a-quarter pounds; it is no wonder that they were alarmed at the prospect of counterparts in Germany thinking along the same lines. 2.1 centimeters is less than the radius of a tennis ball.

The modern value for the density of ^{235}U is about 18.7 gr cm^{-3} ($n_{22} = 4.792$), and that for the fission cross section is about 1.24 barns. These figures give a critical radius (with the factor of 0.8) of 13.46 cm, for a mass of just over 150 kg. Since the critical mass is inversely proportional to the cube of the cross section, overestimating the cross section by a factor of $\sim$8 results in underestimating the mass by a factor of over 500, aside from any change in the adopted density and the effects of scattering.

Had Frisch and Peierls had a sense of the true cross section, we have to wonder if the memorandum would have ever been prepared and how history might have been different. Sometimes an erroneous estimate can propel events more powerfully than an accurate one.

4.5.2 The Fission Cross Section

Frisch and Peierls' choice of 10 barns for the fission cross section of ^{235}U is likely the most controversial aspect of the memorandum among physicists.

Various arguments for and against this value and how F & P might have come to it have been advanced. These are discussed in this section and summarized in Table 4.2. It has to be emphasized, however, this this is all after-the fact speculation; we will never know what sources or whom they might have consulted.

This subsection will refer to comparisons between 1940-era fission cross sections and modern values for ^{235}U and ^{238}U for both fast and thermal neutrons. The latter are summarized in Table 4.1. "Fast" here refers to the cross section averaged across the energy spectrum of fission-liberated neutrons, and "thermal" to neutrons of energy 0.025 eV.

We can begin with the work of Anderson et al. (1939a), published in the March 1, 1939 edition of the *Physical Review*. These authors estimated the fission cross section for fast neutrons on natural uranium to be about 0.1 bn; this is about one-third of the modern value, which is of course dominated by the high abundance ^{238}U. However, with a ratio of $\sim$140:1 for the abundance of ^{238}U to ^{235}U and going on Bohr's prediction that only ^{235}U should undergo fission, this would extrapolate to $\sigma_f \sim 14$ bn for ^{235}U, depending on how the maximum energy of Anderson et al.'s neutrons compared to the threshold for ^{238}U; their paper did not quantify this. If the neutrons were below the fission threshold for ^{238}U, that is, if all fissions were due to ^{235}U, Frisch and Peierls' 10-barn figure is quite plausible.

Supporting evidence of a sort can be inferred from results published by Goldstein et al. (1939) of the Radium Institute in Paris in the July 29, 1939 edition of *Nature*; Frisch and Peierls are likely to have seen this given that Nature was the premier

Table 4.2 ^{235}U fission cross section sources

Reference	Value (bn) /expression	Comments
Anderson et al. (1939a)	0.1	Fast neutron fission; natural U Extrapolates to $\sim$14 bn for ^{235}U
Goldstein (1939)	11.2	Total cross section, natural U Broad neutron energy spectrum
Ladenburg et al. (1939)	0.5	2.4 MeV neutrons; dominated by ^{238}U
Tuve in Bohr and Wheeler (1939)	0.003 (0.6 MeV); 0.012 (1 MeV)	Extrapolates to 0.4 and 1.7 bn for ^{235}U
Nuclear area; Eq. (4.8)	$\sim$2.3	Maximum for "fast" neutrons on $A = 235$
de Broglie wavelength	$\sim 6.43/E_{MeV}$	$\sim$13 bn for $E_{MeV} = 0.5$
Peierls in Dalitz (1997)	3	Origin unknown
Chadwick (2021)	1.24	Modern value

British journal for breaking results. This group reported a *total* cross section of 11.2 bn for fast neutrons on natural uranium, that is, the total for scattering plus fission as opposed to fission alone as reported by Anderson et al. If Frisch and Peierls' assumptions that there is no scattering and that essentially every fast neutron would cause a fission were valid, the 11.2 bn figure would support the inference of 10 bn from the Anderson et al. paper. This would leave little or no excess cross section available for scattering, as they assumed.

There are, however, some serious caveats with this argument. The Paris group used neutrons emitted from a polonium-beryllium source, that is, where alpha particles emitted in the natural decay of polonium strike a beryllium target as described in Chap. 1. If polonium alphas strike beryllium nuclei head-on, the resulting neutrons have energies of ~ 11 MeV, so, at face value, it sounds as if only fast neutrons would be created. However, Goldstein et al. remark that the spectrum of Po-Be neutrons "... has been considered to contain 50 per cent of neutrons ...less than 10^5 eV." Their energy spectrum could thus have hardly been restricted to only fast neutrons; they must have been detecting reactions from both isotopes across a broad swath of energy. In comparison to their result, the modern value for the total cross section of ^{238}U when averaged across the energy spectrum of fission-liberated neutrons is about 7.8 bn; the corresponding figure for thermal neutrons is about 12 bn, not far from their value of 11.2. Anderson et al. filtered their neutrons into slow and fast groups, but the details of the spectrum were not quantified. At best we have a shaky consistency argument here.

Just two weeks before the Goldstein et al. paper appeared, Ladenburg et al. (1939) and collaborators at Princeton University published an estimate of $\sigma_f \sim 0.5$ bn for uranium struck by 2.4-MeV neutrons. This was reported in Bohr and Wheeler's paper on the mechanism of fission, which Frisch and Peierls must have seen. 2.4 MeV is well above Bohr and Wheeler's estimate of the threshold energy for ^{238}U fission, so most of the 0.5 b must pertain to that isotope; there is no way to extract a value for ^{235}U alone. The enormous abundance of ^{238}U could conceivably have "hidden" a 10-b value for ^{235}U, but this would be entering the realm of pure conjecture. The Ladenburg et al. work is mentioned here only for sake of completeness.

More relevant in Bohr and Wheeler's paper is that they reported unpublished results from Merle Tuve and his collaborators at the Carnegie Institution of Washington which indicated fission cross-sections for natural uranium of 0.003 and 0.012 bn at energies of 0.6 and 1 MeV. With the 140:1 abundance ratio and presuming that these figures pertain to pure ^{235}U (neutron energies below the ^{238}U threshold), they imply ~ 0.4 and 1.7 bn for pure ^{235}U; these figures do not differ wildly from the modern values at these energies, ~ 1.1 and 1.2 bn, respectively. If Frisch and Peierls saw this, did they feel that the measurements were as yet too uncertain on which to base a critical mass?

Perhaps given the dearth of and uncertainties surrounding experimental data at the time, Frisch and Peierls might simply have decided to fall back on a theoretical estimate, knowing that further laboratory work would be undertaken. Two theoretical arguments would have indicated opting for a higher value, as described in what follows.

4.5 Paragraph [10]: The Critical Radius and Mass

Frisch and Peierls would surely have been aware of Bohr's conjecture that fission cross-sections for fast neutrons should not exceed the geometrical cross-section of the target nucleus. The idea behind this is that the de Broglie wavelength of a particle - here, a bombarding neutron - decreases as its speed increases; this is quantified below. This has the consequence that if a neutron is moving fast enough, the size of its effective quantum-mechanical wave nature becomes so small compared to a target nucleus that it can be treated as an incoming point particle; the interaction area can then be no larger than the physical cross-sectional area of the nucleus.

From scattering experiments, it was known that nuclear radii behaved as $r \sim r_o A^{1/3}$, where A is the mass number and r_o is an empirical fitting constant. This model assumes that nuclei can be treated as an incompressible collection of nucleons packed tightly together. Today the value of r_o is usually quoted as $r_o \sim 1.2$ fm (fm = femtometer = 10^{-15} m). A slightly larger contemporary 1940 value of $r_o \sim 1.4$ fm is quoted in Bohr and Wheeler's paper (Bethe, 1937), so I will use that here. This gives an area of

$$\sigma_{bn} \sim 0.06158 \, A^{2/3}. \tag{4.8}$$

Exercise: Check this expression. I retain more decimal places than justified so as not to lose accuracy. For $A = 235$, this gives $\sigma \sim 2.34$ bn (modern estimate ~ 1.7). This is larger than the then-going experimental values cited above, but well-shy of F & P's adopted 10 bn.

But there is a loophole in this argument: How fast is "fast"? If what Bohr had in mind was indeed comparing the de Broglie wavelength of a neutron to the size of the nucleus, a larger value of σ can be argued for, as described here.

The de Broglie wavelength is given by

$$\lambda_{deB} = \frac{h}{\sqrt{2Em}}, \tag{4.9}$$

where h is Planck's constant and E is the energy and m the mass of the neutron. For energies in MeV, this becomes

$$\lambda_{deB} = \frac{2.860 \times 10^{-14}}{\sqrt{E_{MeV}}} \text{ m}. \tag{4.10}$$

If we take this to be the effective diameter of an interaction, then the $\pi(\lambda_{deB}/2)^2$ area becomes, in barns,

$$\sigma_{deB} \sim \frac{6.43}{E_{MeV}} \text{ bn}. \tag{4.11}$$

As described in the following section, Frisch and Peierls took fission-liberated neutrons to have $E_{MeV} \sim 0.5$, although the source they may have adopted this from would actually indicate a higher value. But 0.5 MeV gives $\sigma_{deB} \sim 13$ bn, so we are very much back to the 10 bn vicinity. Pearson (2024) indicates that this estimate should in fact be considered to be a lower limit.

There is one final mystery here. In the version of the memorandum printed in Peierls' selected papers (Dalitz & Peierls, 1997), he appended a comment that the assumption of 10 bn was "a stupid slip" and that they should have taken the geometric cross-section for "medium energy neutrons" to be 3 bn. However, there is no indication of where this number might have come from, but using the present-day average energy of fission-liberated neutrons as about 2 MeV, Eq. (4.11) does give $\sigma_{deB} \sim 3.2$ bn. To be fair, however, this comment occurred over 50 years after the fact!

With a cross section of 3 bn, Eq. (4.7) gives a critical radius of about 6.94 cm for Frisch and Peierls' density; the corresponding mass is about 21 kg, about half of the true value.

Further, the simplest explanation might be that 10 barns was no more than a guess. In a 1967 interview, Frisch is quoted as saying that "... I put in a vastly optimistic figure for the fission cross section ..."; Frisch (1967).

4.6 Paragraph [11]: Speed of the Reaction

This paragraph gets to the speed of the neutrons involved in a fast-neutron chain reaction, and the characteristic timescale of the reaction. Concepts involved in this section also play roles in the following one. Unfortunately, further elements of conceptual and numerical confusion also arise here.

Some context for this section will be helpful. In Appendices C and D, it is shown that the time rate of the growth of the neutron population within an exploding bomb core behaves as $e^{(\alpha/\tau_o)t}$. In this expression, α is a dimensionless parameter that arises in the analysis of the neutron diffusion equation. To determine an *initial* value for it, one needs to specify the critical radius R_{co} for the material concerned, the initial radius R_0 of the core, and the number of neutrons per fission ν; one then solves a transcendental equation involving a cotangent; see Eq. (C.16) and following. τ_o is the neutron travel-time-between fissions of Eq. (4.4) above:

$$\tau_o = \frac{\lambda_{cm}}{\langle v \rangle} = \frac{100}{\langle v \rangle \sigma_{bn} n_{22}}. \tag{4.12}$$

α and τ_o are often combined as $\tau = \tau_o/\alpha$, which lets one write the neutron population growth more compactly as $e^{(t/\tau)}$. τ is known as the characteristic timescale for the neutron growth.

The complexities about to arise stem from the fact that both α and τ_o are functions of time. As the core expands, its radius increases from its initial value R_0, so the transcendental equation for α has to be solved anew at each time-step modeled. The result is that α declines from its initial value to zero to the moment of criticality shutdown. In the case of τ_o, the culprit is the density n_{22} in Eq. (4.12): As the core expands, its density decreases, so τ_o increases. Overall, $\tau = \tau_o/\alpha$ increases since the numerator increases while the denominator decreases. The memorandum is phrased

4.6 Paragraph [11]: Speed of the Reaction

in such a way that Frisch and Peierls argue that α can be approximated as unity, so that their τ becomes equivalent to τ_o, but there are inconsistencies in their numbers. This is described in what follows.

Frisch and Peierls took fission-liberated neutrons to have speeds of $\sim 10^9$ cm s^{-1}. From Eq. (4.5), this would demand $E_{MeV} \sim 0.52$. When they were preparing the memorandum, however, there was little information available on the speed spectrum of fission neutrons, but the Zinn and Szilard (1939) paper cited in Sect. 4.2 reported some measurements based on the energetics of recoils of hydrogen atoms which had been struck by fission neutrons. Neutrons of energies up to ~ 3.5 MeV were recorded, with a peak in the distribution at ~ 0.5 MeV. The high-energy tail of the distribution might have suggested a higher average value, but 0.5 MeV is evidently what F & P adopted.

They then remark that "The neutrons emitted in the fission have velocities of about 10^9 cm/s and they have to travel 2.6 cm before hitting a uranium nucleus." This is a simple application of Eq. (4.3) for the mean free path. As described in Sect. 4.5.1, F & P's adopted density for uranium gives a number density of $n_{22} = 3.844$. With their adopted cross section $\sigma_{bn} = 10$, we get

$$\lambda_{cm} = \frac{100}{\sigma_{bn} n_{22}} = \frac{100}{(10)(3.844)} = 2.60 \text{ cm}.$$

Equation (4.12) then gives the timescale as

$$\tau_o = \frac{\lambda_{cm}}{\langle v \rangle} = \frac{2.6 \text{ cm}}{10^9 \text{ cm/s}} = 2.6 \times 10^{-9} \text{ s}.$$

Reflect on this for a moment: On average, a neutron will have a lifetime of only about 2.6 ns before being consumed in causing another fission!

There then follows the statement that "For a sphere well above the critical size the loss through neutron escape would be small, so we may assume that each neutron, after a life of 2.6×10^{-9} s, produces fission, giving birth to two neutrons. In the expression $e^{t/\tau}$ for the increase of neutron density with time, τ would be about 4×10^{-9} s ..."

There are two related issues here: The comment regarding a sphere "well above the critical size", and how 2.6 ns morphs into 4 ns. We will need two relationships to examine these. The first is that between τ and τ_o above:

$$\tau = \frac{\tau_o}{\alpha}. \tag{4.13}$$

The second is derived in Appendix C, where it is shown that an *approximate* expression for the initial value of α is [see Eq. (C.26)]

$$\alpha \sim (\nu - 1)\left[1 - \left(\frac{R_{co}}{R_0}\right)^2\right]. \tag{4.14}$$

Following F & P's argument, if we assume a large core, then $R_0 \gg R_{co}$, and the square bracket reduces to unity. If we further take $\nu \sim 2$ as they argue, then α itself reduces to unity, leaving $\tau = \tau_0$ so that the subsequent focus can remain solely on τ_o. In practice, the problem with this is that when it comes to numerical examples, F & P consider cores of quite finite radii and go back to their value of $\nu = 2.3$.

What of their $\tau = 4$ ns? From Eq. (4.13), $\tau = 4$ and $\tau_o = 2.6$ give $\alpha = 0.65$. However, for their numerical example in the following section, F & P took $R_0 = 2R_{co}$, that is, a core of two critical radii, equivalent to eight critical masses. With this and $\nu = 2.3$, Eq. (4.14) gives $\alpha = 0.975$. How can we reconcile 0.65 and 0.975?

Speculation: If F & P did invoke an argument along this line and they inadvertently dropped the square on $(R_{co}/R_0)^2$ in Eq. (4.14), they would indeed have arrived at $\alpha = 0.65$. If they did underestimate α, this would have the consequence of overestimating τ, which results in underestimating the bomb yield as it depends on the inverse-square of τ; see Eq. (4.17) below.

Another speculation on the origin of $\tau = 4$ ns is offered here. As described following Eq. (4.12) above, τ_o is inversely proportional to the density of the fissile material. As the explosion proceeds and the core expands, its density will drop and τ_o will increase, aside from any effect involving α. Might F & P have simply adopted a larger value for τ_o to reflect this effect? We can make an estimate as follows.

Refer to Appendix D, where it is shown that for a core of initial radius R_0, criticality shuts down when the core has expanded to a radius R_{shut} given by

$$R_{shut} = R_0 \sqrt{\frac{R_0}{R_{co}}}, \qquad (4.15)$$

where R_{co} is again the critical radius for the material concerned.

Write this as $R_{shut}/R_0 = \sqrt{R_0/R_{co}}$. Now, for a fixed mass of material, density is proportional to R^{-3}, so we can say that at criticality shutdown,

$$\frac{\rho_{shut}}{\rho_o} = \left(\frac{R_0}{R_{shut}}\right)^{-3} = \left(\frac{R_{co}}{R_0}\right)^{3/2}, \qquad (4.16)$$

where ρ_o is the initial density. For their (later) choice of $R_0 = 2R_{co}$, this gives $\rho_{shut}/\rho_o = 1/2^{3/2} = 0.354$. At shutdown we would then have $\tau_o = 2.6/0.354 = 7.35$ ns. The average of 2.6 and 7.35 is about 4.98; given that the expansion is slower at first, one can argue that the effective time-average will be somewhat less than the arithmetic average. But again, this is speculation. However F & P arrived at $\tau = 4$ ns, we adopt this value to follow their calculations. Ultimately, the simplest explanation may be that in an exponential growth scenario, if a population grows by a factor of 2 in 2.6 ns, it will grow by a factor of "e" in about 3.8 ns.

4.7 Paragraph [12]: The Yield

This may be the most consequential paragraph of the entire memorandum: A quantitative estimate of the energy yield of a bomb with core mass M and other symbols as defined above. A (possible) derivation of Frisch and Peierls' yield formula can be found in Appendix D. The expression is

$$E = 0.2\, M \left(\frac{R_0}{\tau}\right)^2 \left[\sqrt{\frac{R_0}{R_{co}}} - 1\right]. \tag{4.17}$$

As described above, F & P adopted $\tau = 4 \times 10^{-9}$ s and $R_{co} = 2.1$ cm. As an example, they work with a core of initial radius $R_0 = 4.2$ cm, that is, of two critical radii. As explained in Appendix D, the yield formula assumes a core not much bigger than critical, so 4.2 cm is definitely on the large side. For their density of 15 gr cm^{-3}, the corresponding mass is 4655 grams; they wrote this as 4700 but I will use the more precise value for consistency. With mass in grams, radii in centmeters, and time in seconds, the energy will emerge in ergs:

$$\begin{aligned}
E &= 0.2\, M \left(\frac{R_0}{\tau}\right)^2 \left[\sqrt{\frac{R_0}{R_{co}}} - 1\right] \\
&= 0.2\,(4655) \left(\frac{4.2}{4 \times 10^{-9}}\right)^2 \left[\sqrt{2} - 1\right] = 4.25 \times 10^{20} \text{ ergs} \\
&= 4.25 \times 10^{13} \text{ Joules} \sim 10.1 \text{ kt}.
\end{aligned}$$

Frisch and Peierls quote 4×10^{20} ergs. Curiously, the Hiroshima bomb is estimated to have had a yield of about 12–13 kt, but this is not a fair comparison: The Little Boy bomb was heavily tamped and utilized about 64 kg of enriched uranium.

This 10 kt represents an efficiency of about 10%, as F & P claim. With an atomic weight of 235 grams per mole, 1 kg of ^{235}U contains about 4.255 moles of nuclei, that is, about 2.563×10^{24} nuclei. If each fission liberates 170 MeV of energy, our 1 kg is equivalent to 6.980×10^{13} Joules, or 16.62 kt. A 4.655 kg core then has a theoretical yield of 77.4 kt; the efficiency is then $10.1/77.4 \sim 13\%$. The actual efficiency of the Hiroshima bomb was about $1 - 2\%$.

An aside argument: In the qualitative part of the memorandum, F & P estimated that the explosion would, "for an instant" produce a temperature comparable to that of the interior of the Sun. This can be understood with the numbers developed here. From thermodynamics, the total "internal energy" of an ideal gas is given by $E = 3NkT/2$, where N is the number of particles involved and k is Boltzmann's constant, 1.381×10^{-23} Joules per Kelvin. For the complete fission of a 1 kg core, the 2.563×10^{24} nuclei will each fission into two product nuclei, giving $N = 5.126 \times 10^{24}$. Then with $E = 6.980 \times 10^{13}$ Joules, you will find $T \sim 6.6 \times 10^{11}$ Kelvins. In

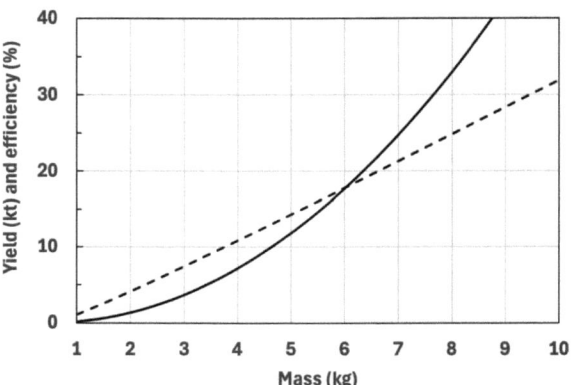

Fig. 4.1 Yield in kilotons (solid curve) and efficiency (dashed curve) corresponding to Frisch and Peierls' yield formula. The vertical scale does dual duty as both kilotons and percentage. $\tau = 4 \times 10^{-9}$ s, $\rho = 15 \,\text{gr cm}^{-3}$, $R_{co} = 2.1$ cm, and 170 MeV per fission

contrast, the temperature in the core of the Sun is estimated to be a meager $\sim 1.5 \times 10^7$ K, so their estimate was very conservative!

Figure 4.1 shows the yield in kilotons (solid curve) and efficiency (dashed curve) corresponding to Eq. (4.17). The vertical scale reads as both kilotons and percentage; $\tau = 4 \times 10^{-9}$ s, $\rho = 15 \,\text{gr cm}^{-3}$, $R_{co} = 2.1$ cm, and energy 170 MeV per fission. Given the confusion surrounding τ described in the previous section, however, these results need to be taken with a grain of salt.

If τ is kept fixed at 4 ns, the entire fission energy is liberated for a core of mass ~ 32 kg as F & P state, but this would represent a circumstance well beyond the range of the approximations underlying the yield formula. It is interesting to note that, as described in Appendix D, the yield is independent of energy per fission, but efficiency is inversely proportional to the energy per fission.

Frisch and Peierls now make a remark that leads to another numerical mystery: "For small radii the efficiency falls off even faster than indicated by formula (1) because τ goes up as R_0 approaches R_{co}. The energy liberated by a 5 kg bomb would be equivalent to that of several thousand tons of dynamite, while that of a 1 kg bomb, though about 500 times less, would still be formidable."

When computing the efficiency, the yield is divided by the product of the mass of the core and the factor of 16.62 kilotons per kilogram derived above. The mass cancels, and the efficiency behaves as $(R_0/\tau)^2$ times the square-bracketed factor in Eq. (4.17). If the core radius is "small", we can then approximate $(R_0/\tau)^2 \rightarrow (R_{co}/\tau)^2$. The square-bracketed factor will then also approach zero as they claim, so the efficiency will be very poor in such a case. As to the comment that τ goes up as R_0 approaches R_{co}, this can be understood on the basis that, for a core that is just slightly larger than critical, the density will alter but little before the explosion shuts down. This means that $\tau = \tau_o/\alpha$ will undergo little change from any alteration of τ_o, but for such a core α will be intrinsically small to begin with, which will lead to a large value of τ than would be the case for a larger core.

So far as the comparison of a 5-kg core to a 1-kg core goes, if τ is left at 4 ns, the yields are respectively about 11.9 and 0.18 kt, for a ratio of about 65. If the yield ratio is assumed to be ∼500 as claimed, Eq. (4.17) can be used to determine that τ for the 1-kg core must be about 11 ns if the value of 4 ns is retained for the 5-kg core. This author has made various attempts to recover the factor of 500 by positing various assumptions regarding how F & P might have treated α and τ_o such that the results of the 4.7-kg example are not disturbed, but without success. With no clues as to what they might have done, their treatment of this example seems destined to remain frustratingly opaque.

4.8 Paragraphs [13]–[15]: Triggering and Predetonation

In these three qualitative paragraphs, Frisch and Peierls consider aspects of triggering a nuclear weapon, emphasizing that sub-critical components would have to be kept separated until it is desired to detonate the bomb, at which time the assembly would need to be executed as quickly as possible to avoid neutrons originating "from the action of uranium alpha rays on light-element impurities." In the qualitative part of the memorandum as described in Chap. 3, they estimated the probability of a bomb failing in this way to be about 1 in 100; a very rough estimate of this is developed in what follows.

In considering the possibility of an (α, n)-initiated predetonation, the issue is to minimize the extent of any light-element impurities in the fissile material; beryllium and aluminum are particularly problematic in this regard. Chemical processing of the fissile material will inevitably introduce some level of impurities, and since uranium (and plutonium) is a natural alpha-emitter, the possibility of a predetonation is always unavoidable; a stray neutron could initiate a premature reaction as soon as first criticality is achieved. The rates of alpha-emission per gram of material are fixed by nature, so minimizing the possibility of a predetonation during core assembly means minimizing the level of impurities and making the assembly time as short as possible. Quaintly, F & P suggested that springs might be used to bring the sub-critical pieces together; at Los Alamos, chemical explosives would be used to propel the pieces together over time intervals on the order of 100 microseconds once the pieces began to mate.

The theory of radioactive decay provides an expression for the rate of alpha-decays per second per kilogram of fissile material in terms of its atomic weight A and half-life $t_{1/2}$ (seconds) for alpha decay. This is

$$R_\alpha = 10^3 \left(\frac{N_A}{A}\right)\left(\frac{\ln 2}{t_{1/2}}\right), \quad (4.18)$$

where N_A is Avogadro's number. The factor of 10^3 arises on converting rates per gram to rates per kilogram.

Numbers for ^{235}U and ^{239}Pu appear in Table 4.3. We will be concerned only with the former as Frisch and Peierls had no knowledge of plutonium, but the latter is

Table 4.3 Alpha-decay data

Isotope	Half-life (years)	Half-life (seconds)	Alpha-decay rate $kg^{-1} s^{-1}$
^{235}U	7.04×10^8	2.22×10^{16}	8.0×10^7
^{239}Pu	24,100	7.61×10^{11}	2.3×10^{12}

included to point out its much greater rate of decay. In the case of ^{235}U, 80 million decays per second for a kilogram of material is a lot, but remember that each cubic centimeter of material contains over 10^{22} nuclei; the chance of any one nucleus decaying in a given one-second interval is quite remote. So far as the (α, n) problem goes, however, the large decay rate is mitigated by two factors: The so-called *yield* (y) of such reactions, and the fact that the any impurity will presumably have been reduced to a low level: If you are going to invest in making a bomb, don't cheap out on the chemical processing. More sophisticated analyses than the one developed here take into account more detail on the nature of the impurity involved; the intent here is an order-of-magnitude estimate.

As alpha particles travel through material, they quickly lose energy by ionizing atoms that they pass close to; in a solid material they will be brought to a stop in a few millimeters. The chance of an alpha actually striking another nucleus during this time is also quite remote: Atoms are mostly empty space. Nuclear experimentalists quantify these considerations through the yield factor. Experimentally, most light-element (α, n) reactions have yields on the order of $y \sim 10^{-5} - 10^{-4}$; to be conservative, I will work with the upper limit. This means that only 1 in 10,000 alphas actually strikes another nucleus as they travel through a lump of material. If we are working with one kilogram of ^{235}U, the 80 million decays per second then reduces to 8,000 alpha-nucleus interactions per second.

With good chemical processing, you should be able to reduce the level of an impurity to on the order of parts-per-million, that is, of every million nuclei in the core, only one should be of an impurity. (If there is more than one impurity, the effects will be additive; I stick to considering one here.) Of the 8,000 alpha-nucleus collisions per second, then only 10^{-6} of them, or some 0.008 per second will have the misfortune to be an alpha-impurity strike which will liberate a neutron. If the subcritical core pieces are assembled by some simple mechanical arrangement as F & P had in mind, let us suppose that there is a danger period of one-tenth of a second between when a critical mass is present in the partially-assembled system and assembly is complete. This takes us down to 0.0008 neutrons per kilogram during the danger period. Some of these will escape the core without causing a fission, but I will again err on the side of caution; also, the 0.1 s danger time is a wild overestimate, as evidenced by the discussion of the 100-microsecond timescale related above. Even if the bomb core should have a mass of 10 kg (more than F & P anticipated), we are back to only 0.008 neutrons during the critical time, or about a 1% chance (0.01) of initiating the chain reaction prematurely. If anything, their estimate of a failure rate of 1 in 100 was on the pessimistic side.

In a sense, this analysis is a cheat in that Frisch and Peierls did not know the alpha-decay properties of ^{235}U, as a sample of that isotope had not yet been separated in any appreciable quantity. But the half-life for natural U (dominated by ^{238}U) was known to be about 4.5×10^9 years, about 6.5 times *greater* than that for ^{235}U. They would have calculated a more demanding level of purity than that used here, but no so much as to indicate that a predetonation could not be reduced to an acceptable level. A further then-unknown property of uranium is that it also fissions spontaneously, although the half-life for this process is very great, about 10^{19} years for ^{235}U. In this case, however, neutrons are emitted directly; the half-life is not "buffered" by any mitigating yield factor.

Presciently, F & P remark that "By experimenting with spheres of gradually increasing size and measuring the number of neutrons emerging from them under a known neutron bombardment, one could accurately determine the critical size, without any danger of a premature explosion." This is exactly what would be done in "criticality" experiments a few years later at Los Alamos. In addition to Otto Frisch's "dragon" experiments described in Chap. 3, assemblies of varying amounts of ^{235}U (and also plutonium) were arranged to approximate critical masses.

4.9 Paragraphs [16]–[18]: Thermal Diffusion and Isotope Separation

These three paragraphs address the question of how to isolate kilogram-level quantities of ^{235}U, focusing on the method of thermal diffusion. The particular method Frisch and Peierls advocated would come to be known in the Manhattan Project as "liquid" thermal diffusion as opposed to "gaseous" thermal diffusion, which would be used in a much larger facility.

As described in the memorandum, the principle of liquid diffusion is that if a fluid (which can mean a gas or liquid) comprising atoms/molecules containing two isotopes of an element is pumped into a space which is bounded by walls of very different temperatures, molecules of the lighter isotope will tend to accumulate toward the hotter wall while those of the heavier isotope accumulate toward the cooler wall. Fluid containing the lighter isotope will be of lower density and consequently rise by convection and so preferentially collect toward the top of the space, while that containing the heavier isotope will fall. Competition between this thermally-induced process and the ordinary diffusion of the isotopes through each other will lead, after some hours or days, to an equilibrium between the two processes.

The theory of thermal diffusion was first developed by David Enskog in Sweden in 1911 and Sydney Chapman in England (1916); experimental proof was established by Chapman and F.W. Dootson in 1917. In Germany, Klaus Clusius and Gerhard Dickel first used a "column" approach in 1938 by placing a hot wire along the central axis of a vertical tube, and achieved a small enrichment of neon isotopes. Soon thereafter, Arthur Bramley and Keith Brewer of the U.S. Department of Agriculture conceived the idea of using two concentric tubes held at different temperatures. Bramley and Brewer used steam to heat the inner tube and water to cool the outer one, while injecting the fluid to be processed into a narrow annulus between them. This is the technique that would be used at Oak Ridge, Tennessee. Figure 4.2 shows a diagram of Manhattan Project thermal diffusion column.

The metric of performance for such a column is its so-called separation factor, which specifies its enrichment capability. For example, if a column has a separation factor of 1.2 and natural uranium is used, then, after processing, the percentage of ^{235}U will be $0.720\%(1.2) = 0.864\%$. 1.2 was the value for the Manhattan Project's columns, which were designated as the S-50 facility. Material so processed is harvested from the top of a column and sent on to another such stage. Since only a small fraction of the material in a column is taken off in any harvesting, the optimal arrangement is to have a great number of "first stage" columns operating in parallel to produce material which is harvested and fed into a smaller number of second-stage columns and so on until the desired enrichment is reached. In theory, a succession of 27 stages would bring 0.72% material to 90% material, but, given the over 100-to-1 isotope ratio for uranium, vast quantities of feed material are required.

The theory of diffusion is immensely complex; a colleague of this author who is familiar with some of the details admits to being largely baffled by it. Ultimately,

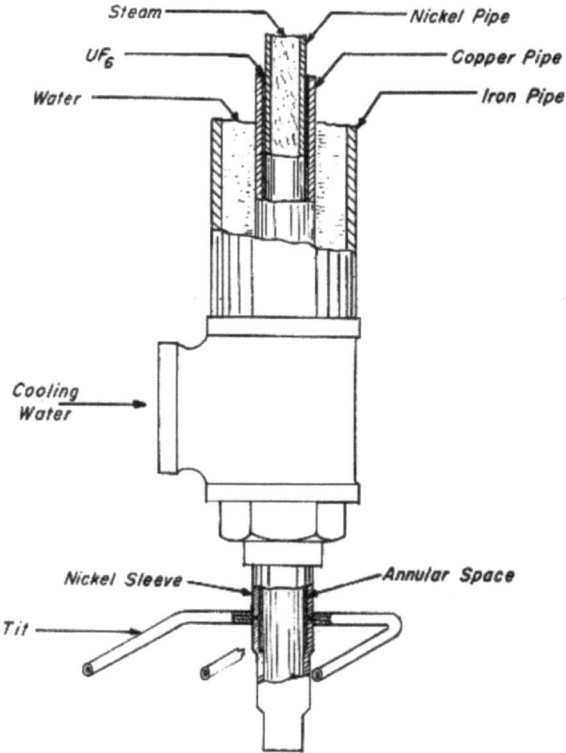

Fig. 4.2 Sectional view of a Manhattan Project 48-foot thermal diffusion column. Uranium hexafluoride (UF_6) of natural isotopic uranium abundance is driven into the narrow annular space (0.25 mm wide) between the nickel and copper pipes; the nickel pipes were 1.25 inches outside diameter. At the top and bottom of each tube, three small projecting "tits" provided access to the annular space for supply and withdrawal of material. *Source* Reed (2011) with permission of the author

the separation factor and production rate depends on the temperature difference between the columns, the width of the annulus, and the length of the columns. Frisch and Peierls give no details as to how they estimated that tubes of length 150 cm would produce a separation factor of 1.4, but this number seems overly optimistic: the Manhattan Project columns were 48 *feet* in length. They also estimated that 100,000 columns would produce 100 grams of 90% ^{235}U per day. A factor in such an operation is producing the tremendous amounts of steam needed to heat the inner columns; Manhattan's S-50 plant had its own dedicated 238-megawatt powerhouse for this purpose. While Frisch and Peierls envisioned a thermal diffusion plant which would produce bomb-grade material by itself, the S-50 plant was configured with 2,142 columns operating in parallel to provide a large quantity of slightly enriched ^{235}U as feed for other enrichment facilities. By September 1945, S-50 had produced 56,500 pounds of 0.86%-enriched U; this would represent some 220 kg of ^{235}U, enough for about four Hiroshima Little Boy bombs. For a history of the S-50 facility, the author humbly recommends Reed (2011).

4.10 Paragraphs [19]–[24]: Radiation Effects

Nuclear weapons cause tremendous destruction by the shock waves they create, as well as liberating thermal radiation intense enough to incinerate structures and living beings unlucky enough to be in their vicinity; fires can be ignited for miles surrounding the detonation. But Frisch and Peierls were more troubled by the intense radioactive fallout that would be generated, a feature no conventional explosive creates, no matter how powerful. They devote six paragraphs to this issue, making some rough estimates of radiation effects. In examining their numbers, I will rely here more on modern-day information than was done in the preceding section: Morbid as it may seem, hundreds of postwar nuclear tests have yielded detailed information on the fallout effects of nuclear weapons; two good sources are Glasstone and Dolan (1977), and, more compactly, Broyles (1982). Frisch and Peierls' estimates were certainly within an order of magnitude of what would be derived from modern numbers.

Perhaps the most stunning comment in the memorandum is that "Even one day after the explosion the radiation will correspond to a power expenditure of the order of 1000 KW, or to the radiation of a hundred tons of radium." How can the fission of, say, a kilogram of uranium lead to such intense radioactivity?

Some historical background will be helpful here. Radium was first isolated in late 1898 by Marie and Pierre Curie, and recognized as an intense alpha-emitter. The rate of decays of any material of atomic weight A grams per mole and half-life $t_{1/2}$ can be computed with Eq. (4.18); traditionally, rates per *gram* of material were calculated, so just drop the factor of 10^3 from the equation. A single gram of freshly-isolated radium-226 (^{226}Ra) has a decay rate of 3.7×10^{10} per second, which corresponds to a half-life of about 1599 years. This is much shorter than the half-life for ^{235}U (see Table 4.3), which, in comparison, could be considered a rather gentle alpha-emitter.

This decay rate of ^{226}Ra is now taken to define one *Curie* of radioactivity. A Curie is a *lot* of radioactivity; household smoke detectors typically contain a single microcurie of a radioactive substance used to ionize air for a few centimeters around the detector as an aid in detecting smoke particles.

Uranium fission can happen in numerous ways, giving rise to hundreds of isotopes of dozens of elements. The half-lives of fission products range from fractions of a second to years, but the majority are less than a day. This is the reason for the intensity of nuclear fallout: A single gram of a 1-day fission product (of the same atomic weight) is equivalent in its radioactivity to some 584,000 grams of ^{226}Ra (1,600 years = 584,000 days). This is 584 kg of ^{226}Ra, or over half a metric ton (1 metric ton = 1,000 kg, often written as "tonne" to distinguish it from the British 2,000-lb ton). A gram of a 1-minute half-life product is similarly equivalent to over 900 British tons of radium.

As a specific example, the fission product produced most abundantly in uranium fission is strontium-97 ($^{97}_{38}$Sr), which arises in about 0.86% of fissions. This has a half-life of a mere 0.43 s, making a single gram of this material equivalent to about 270 billion grams, or 270,000 tonnes of radium! From Eq. (4.18), the calculation of this goes as (drop the factor of 10^3 to deal with rates per gram of material)

$$R_\alpha = \left(\frac{N_A}{A}\right)\left(\frac{\ln 2}{t_{1/2}}\right) = \left(\frac{6.022 \times 10^{23}}{97}\right)\left(\frac{\ln 2}{0.43}\right) \quad (4.19)$$
$$= 1.001 \times 10^{22} \text{ decays per second per gram.}$$

With one Cuire as 3.7×10^{10} decays per second, this is equivalent to 2.71×10^{11} Curies, or about 270 billion grams, as claimed. This activity will obviously decay very promptly, but makes the point that it is not at all difficult to get into hundreds-of-tons figures.

For comparison with Frisch and Peierls, it is useful to know the radioactivity of 100 tons of radium, and the corresponding power represented by the decays. If they had in mind a 2,000-pound ton, this is equivalent to ~90,700 kg, or 9.07×10^7 grams, that is, 90.7 million Curies, or about 3.36×10^{18} decays per second. Each radium alpha-decay liberates 4.78 MeV of energy, so we have a power of

$$\text{Power} = \left(3.36 \times 10^{18} \frac{\text{decay}}{\text{s}}\right)\left(4.78 \frac{\text{MeV}}{\text{decay}}\right)\left(1.602 \times 10^{-13} \frac{\text{Joule}}{\text{MeV}}\right)$$
$$= 2.57 \times 10^6 \frac{\text{J}}{\text{s}} = 2570 \text{ kW.}$$

F & P's claim 1000 kW for 100 tons of radium was an *under*estimate.

To somewhat check the consistency of F & P's claim that the radiation after one day would still amount to that equivalent to 100 tons of radium, two modern rules of thumb are helpful. First, nuclear weapons engineers divide consideration of the radiation released in an explosion into two time periods: "prompt," generally defined as within one minute of the explosion, and "residual" beyond that time. For the

4.10 Paragraphs [19]–[24]: Radiation Effects

latter, the collective effects of the decay of direct fission products and their decay products gives an effective decay rate that is approximately inversely proportional to the elapsed time to the power 1.2; F & P approximated this as simply inversely proportional to time. This means that if the decay rate is DR_o at some time t_o, then the decay rate at a later time t will be given by

$$DR(t) = DR_o \left(\frac{t_o}{t}\right)^{1.2}. \tag{4.20}$$

The second rule is that at one minute after the explosion, the decay rate is about 30 billion Curies for each kiloton of yield.

Now consider F & P's 10-kt example, for which the initial decay rate will be 300 billion Curies. If we take $t_o = 1$ minute in Eq. (4.20) and setting 1 day = 1440 min, we have

$$DR(1 \text{ day}) = \left(3 \times 10^{11} \text{ Ci}\right) \left(\frac{1}{1440}\right)^{1.2} \sim 4.87 \times 10^7 \text{ Ci},$$

that is, about 49 million Curies, or just over half of what was calculated above for 100 tons of radium. By modern standards they were about a factor of two too great, but this is not a bad estimate considering that no even small-scale nuclear explosion had been carried out at the time. After a week, this will have decayed by a factor of 10, but one is still dealing with a staggering amount of radioactivity.

As F & P remarked, it is difficult to estimate how this translates into effects on humans, given the many contingencies involved: Is the bomb detonated near the ground or aloft? Does wind carry much of the radioactive material away? Geographical factors can concentrate fallout in particular areas. Are you out in the open, or can you take shelter in a place where filtered air and uncontaminated food and water are available? How much contaminated air do you breathe in? Can you change to uncontaminated clothing?

We can make a few comments on this human-exposure business. Aside from blast and burn effects, the most serious form of radiation from a weapon is prompt gamma-rays emitted during the explosion; like X-rays, these are highly penetrating. Today, human radiation exposure is measured in units called "rads," which are very roughly equivalent to the "Roentgens" used by Frisch & Peierls. Technically, a rad is equivalent to absorbing 0.01 Joules of energy per kg of body mass. A whole-body dose of less than 25 rads will generally be harmless; 200 rads will cause redness of the skin; 400 rads is considered to be the dose that will cause 50% lethality, and 1000 rads causes 100% lethality, although localized doses for medical purposes can often be higher than these numbers.

Figure 4.3 shows approximate doses for an exposed person as a function of direct range from the explosion (in yards) for bombs of yield 2, 10, and 50 kt; this data is adapted from Glasstone and Dolan (1977). Both scales are base-10 logarithmic. For Frisch & Peierls' 10-kt bomb, anywhere within a range of ~1,450 yards will give a dose exceeding 400 rems (log 400 = 2.60), and anywhere within about 1,200 yards

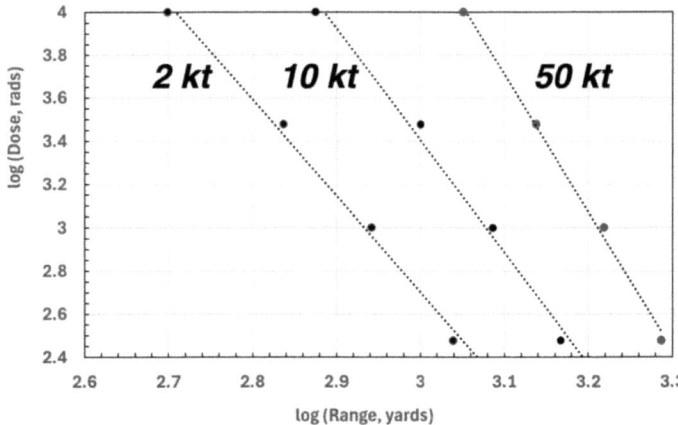

Fig. 4.3 Approximate prompt gamma-ray doses in rads for air-burst fission weapons of yields 2, 10 and 50 kt. Logarithms are base 10. Interpolated from Fig. 8.33a of Glasstone and Dolan (1977). A whole-body dose of 1000 rads will prove lethal in most cases

will receive over 1,000 rads. But if you are that close, the thermal effects will be well-beyond third-degree burns. Best advice: Be either as close as possible or very far away.

References

Anderson, H. L., Booth, E. T., Dunning, J. R., Fermi, E., Glasoe, G. N., & Slack, F. G. (1939a). The Fission of Uranium. *Physical Review, 55*(5), 511–512.
Anderson, H. L., Fermi, E., & Hanstein, H. B. (1939b). Production of neutrons in Uranium bombarded by neutrons. *Physical Review, 55*(8), 797–798.
Anderson, H. L., Fermi, E., & Szilard, L. (1939c). Neutron production and absorption in Uranium. *Physical Review, 56*, 284–286.
Bethe, H. (1937). Nuclear physics B. Nuclear dynamics theoretical. *Reviews of Modern Physics, 9*(2), 69–244.
Bohr, N., & Wheeler, J. A. (1939). The mechanism of nuclear fission. *Physical Review, 56*(5), 426–450.
Broyles, A. A. (1982). Nuclear explosions. *American Journal of Physics, 50*(7), 586–594.
Chadwick, J. (1969). Interview with Charles Weiner. Retrieved April 20, 1969, from https://www.aip.org/history-programs/niels-bohr-library/oral-histories/3974-4
Chadwick, M. (2021). Nuclear science for the Manhattan Project and comparison to today's ENDF data. *Nuclear Technology, 207*(S1), S24–S61.
Clark, R.W. (1961). *The Birth of the bomb: The untold story of Britain's part in the weapon that changed the world.* London: Phoenix House.
Dalitz, R. H., & Peierls, S. R. (1997). *Selected scientific papers of Sir Rudolf Peierls with commentary.* Singapore: World Scientific.
Fermi, E. (1965). *Collected papers, Volume II: United States 1939–1954.* Chicago: University of Chicago Press.

References

Frisch, O.R. (1967). Interview with Charles Weiner, May 3, 1967. https://repository.aip.org/frisch-otto-robert-1967-may-3

Glasstone, S., & Dolan, P. J. (1977). *The effects of nuclear weapons*. Washington: United States Department of Defense and Energy Research and Development Agency.

Goldstein, L., Rogozinski, A., & Walen, R. J. (1939). The scattering by Uranium nuclei of fast neutrons and the possible neutron emission resulting from fission. *Nature, 144*(3639), 201–202.

Gowing, M. (1964). *Britain and atomic energy 1939–1945*. London: St. Martin's Press.

Halban, H., Joliot, V. F., & Kowarski, L. (1939a). Liberation of neutrons in the nuclear explosion of Uranium. *Nature, 143*(3620), 470–471.

Halban, H., Joliot, V. F., Kowarski, L., & Perrin, F. (1939b). Mise en évidence d'une réaction nucléiare en chaîne au seine d'une massse uranifère. *Journal de Physique et Le Radium, 10*(10), 428–429.

Ladenburg, R., Kanner, M. H., Barschall, H. H., & van Voorhis, C. C. (1939). Study of Uranium and Thorium fission produced by fast neutrons of nearly homogeneous energy. *Physical Review, 56*(2), 168–170.

Pearson, J. M. (2024). Comments on the Frisch-Peierls estimate of the critical mass of a uranium fission bomb. *Nuclear Technology, 210*(6), 1078–1082.

Reed, B. C. (2011). Liquid thermal diffusion during the Manhattan Project. *Physics in Perspective, 13*(2), 161–188.

Reed, B. C. (2020b). A simplified treatment of neutron slowing by elastic collisions. *American Journal of Physics, 88*(10), 871–873.

Reed, B. C. (2021b). An inter-country comparison of nuclear pile development during World War II. *European Physical Journal-H, 46*, 15.

Szilard, L., & Zinn, W. H. (1939). Instantaneous emission of fast neutrons in the interaction of slow neutrons with Uranium. *Physical Review, 55*(8), 799–800.

Weart, S. R. (1979). *Scientists in power*. Cambridge, MA: Harvard University Press.

Zinn, W. H., & Szilard, L. (1939). Emission of neutrons by Uranium. *Physical Review, 56*(7), 619–624.

Epilogue

This brief chapter summarizes the contents, impact, and legacy of the Frisch–Peierls memorandum. The thousands of nuclear weapons in existence today trace their lineages to Birmingham in 1940.

The Frisch–Peierls memorandum was by no means a perfect document: There are unexplained assumptions, numerical leaps, and apparent inconsistencies. But what is remarkable is how much they got right: Clear understanding of the distinction between slow and fast-neutron reactions; the concept of a critical mass; assembling a supercritical mass from subcritical parts; the possibility of predetonation; the form of the yield formula; the order of magnitude of the energy release; likely radiological effects; and the strategic/ethical considerations of using such weapons. Their command of the physics and foresight were remarkable. Years of intense research and development as well as unexpected problems lay ahead, but they anticipated most of the major issues.

Nothing written here should be taken to be critical of Frisch and Peierls: They were working in secrecy, and, once aware of the implications of their ideas, likely in haste to get word to responsible authorities. They would have been well aware that numbers and details would clarify as experimental and theoretical work got underway in earnest; the important thing was to get it coordinated and moving. They had to assume that the same ideas would occur to former colleagues back in Germany, and to some extent they already had. Physics was about to move into the sphere of politics and wartime priorities, a vastly more uncertain arena.

The legacy of the memorandum is still very much with us in the form of the thousands of nuclear weapons held by the world's nuclear powers, a situation that would likely have appalled Frisch and Peierls. That many countries are now upgrading and expanding their arsenals would trouble them even more.

© The Editor(s) (if applicable) and The Author(s), under exclusive license to Springer Nature Switzerland AG 2025
B. C. Reed, *The Frisch-Peierls Memorandum*,
SpringerBriefs in History of Science and Technology,
https://doi.org/10.1007/978-3-031-95929-5

Appendix A
The Memorandum: Qualitative Part

This chapter is a reproduction of the qualitative part of the Frisch–Peierls memorandum, adopted from Ronald W. Clark, "Tizard" (Methuen, London, 1965). Paragraph numbers in [square brackets] were not included in the original document.

Memorandum on the Properties of a Radioactive 'Super-Bomb'

[1] The attached detailed report concerns the possibility of constructing a 'super-bomb' which utilises the energy stored in atomic nuclei as a source of energy. The energy liberated in the explosion of such a super-bomb is about the same as that produced by the explosion of 1,000 tons of dynamite. This energy is liberated in a small volume, in which it will, for an instant, produce a temperature comparable to that in the interior of the sun. The blast from such an explosion would destroy life in a wide area. The size of this area is difficult to estimate, but it will probably cover the centre of a big city.

[2] In addition, some part of the energy set free by the bomb goes to produce radioactive substances, and these will emit very powerful and dangerous radiations. The effect of these radiations is greatest immediately after the explosion, but it decays only gradually and even for days after the explosion any person entering the affected area will be killed.

[3] Some of this radioactivity will be carried along with the wind and will spread the contamination; several miles downwind this may kill people.

[4] In order to produce such a bomb it is necessary to treat a few cwt. of uranium by a process which will separate from the uranium its light isotope (U_{235}) of which it contains about 0.7%. Methods for the separation of isotopes have recently been developed. They are slow and they have not until now been applied to uranium, whose chemical properties give rise to technical difficulties. But these difficulties are by no means insuperable. We have not sufficient experience with large-scale chemical plant to give a reliable estimate of the cost, but it is certainly not prohibitive.

[5] It is a property of these super-bombs that there exists a 'critical size' of about one pound. A quantity of the separated uranium isotope that exceeds the critical amount is explosive; yet a quantity less than the critical amount is absolutely safe. The bomb would therefore be manufactured in two (or more) parts, each being less than the critical size, and in transport all danger of a premature explosion would be avoided if these parts were kept at a distance of a few inches from each other. The bomb would be provided with a mechanism that brings the two parts together when the bomb is intended to go off. Once the parts are joined to form a block which exceeds the critical amount, the effect of the penetrating radiation always present in the atmosphere will initiate the explosion within a second or so.

[6] The mechanism which brings the parts of the bomb together must be arranged to work fairly rapidly because of the possibility of the bomb exploding when the critical conditions have just only been reached. In this case the explosion will be far less powerful. It is never possible to exclude this altogether, but one can easily ensure that only, say, one bomb out of 100 will fail in this way, and since in any case the explosion is strong enough to destroy the bomb itself, this point is not serious.

[7] We do not feel competent to discuss the strategic value of such a bomb, but the following conclusions seem certain:

[8] 1. As a weapon, the super-bomb would be practically irresistible. There is no material or structure that could be expected to resist the force of the explosion. If one thinks of using the bomb for breaking through a line of fortifications, it should be kept in mind that the radioactive radiations will prevent anyone from approaching the affected territory for several days; they will equally prevent defenders from reoccupying the affected positions. The advantage would lie with the side which can determine most accurately just when it is safe to re-enter the area; this is likely to be the aggressor, who know the location of the bomb in advance.

[9] 2. Owing to the spread of radioactive substances with the wind, the bomb could probably not be used without killing large numbers of civilians, and this may make it unsuitable as a weapon for use by this country. (Use as a depth charge near a naval base suggests itself, but even there it is likely that it would cause great loss of civilian life by flooding and by the radioactive radiations.)

[10] 3. We have no information that the same idea has also occurred to other scientists but since all the theoretical data bearing on this problem are published, it is quite conceivable that Germany is, in fact, developing this weapon. Whether this is the case is difficult to find out, since the plant for the separation of isotopes need not be of such a size as to attract attention. Information that could be helpful

Appendix A: The Memorandum: Qualitative Part

in this respect would be data about the exploitation of the uranium mines under German control (mainly in Czechoslovakia) and about any recent German purchases of uranium abroad. It is likely that the plant would be controlled by Dr. K. Clusius (Professor of Physical Chemistry in Munich University), the inventor of the best method for separating isotopes, and therefore information as to his whereabouts and status might also give an important clue.

[11] At the same time it is quite possible that nobody in Germany has yet realized that the separation of the uranium isotopes would make the construction of a super-bomb possible. Hence it is of extreme importance to keep this report secret since any rumour about the connection between uranium separation and a super-bomb may set a German scientist thinking along the right lines.

[12] 4. If one works on the assumption that Germany is, or will be, in the possession of this weapon, it must be realized that no shelters are available that would be effective and that could be used on a large scale. The most effective reply would be a counter-threat with a similar bomb. Therefore it seems to us important to start production as soon and as rapidly as possible, even if it is not intended to use the bomb as a means of attack. Since the separation of the necessary amount of uranium is, in the most favourable circumstances, a matter of several months, it would obviously be too late to start production when such a bomb is known to be in the hands of Germany, and the matter seems, therefore, very urgent.

[13] 5. As a measure of precaution, it is important to have detection squads available in order to deal with the radioactive effects of such a bomb. Their task would be to approach the danger zone with measuring instruments, to determine the extent and probable duration of the danger and to prevent people from entering the danger zone. This is vital since the radiations kill instantly only in very strong doses whereas weaker doses produce delayed effects and hence near the edges of the danger zone people would have no warning until it were too late.

[14] For their own protection, the detection squads would enter the danger zone in motor-cars or aeroplanes which are armoured with lead plates, which absorb most of the dangerous radiation. The cabin would have to be hermetically sealed and oxygen carried in cylinders because of the danger from contaminated air.

[15] The detection staff would have to know exactly the greatest dose of radiation to which a human being can be exposed safely for a short time. This safety limit is not at present known with sufficient accuracy and further biological research for this purpose is urgently required.

[16] As regards the reliability of the conclusions outlined above, it may be said that they are not based on direct experiments, since nobody has ever yet built a super-bomb, but they are mostly based on facts which, by recent research in nuclear physics, have been very safely established. The only uncertainty concerns the critical size for the bomb. We are fairly confident that the critical size is roughly a pound or so, but for this estimate we have to rely on certain theoretical ideas which have not been positively confirmed. If the critical size were appreciably larger than we believe it to be, the technical difficulties in the way of constructing the bomb would be enhanced. The point can be definitely settled as soon as a small amount of uranium has been separated, and we think that in view of the importance of the matter immediate steps

should be taken to reach at least this stage; meanwhile it is also possible to carry out certain experiments which, while they cannot settle the question with absolute finality, could, if their result were positive, give strong support to our conclusions.

O. R. Frisch
R. Peierls

The University, Birmingham

Appendix B
The Memorandum: Technical Part

This chapter is a reproduction of the quantitative part of the Frisch–Peierls memorandum. Paragraph numbers in [square brackets] were not included in the original document.

Strictly confidential

On the Construction of a "Super-Bomb", Based on a Nuclear Chain Reaction in Uranium

[1] The possible construction of "super-bombs" based on a nuclear chain reaction in uranium has been discussed a great deal and arguments have been brought forward which seemed to exclude this possibility. We wish here to point out and discuss a possibility which seems to have been overlooked in these earlier discussions.

[2] Uranium consists essentially of two isotopes, U_{238} (99.3%) and U_{235} (0.7%). If a uranium nucleus is hit by a neutron, three processes are possible: (1) scattering, whereby the neutron changes direction and, if its energy is above about 0.1 MeV, loses energy; (2) capture, when the neutron is taken up by the nucleus; and (3) fission, i.e. the nucleus breaks up into two nuclei of comparable size, with the liberation of an energy of about 200 MeV.

[3] The possibility of a chain reaction is given by the fact that neutrons are emitted in the fission and that the number of these neutrons per fission is greater than 1. The most probable value for this figure seems to be 2.3, from two independent determinations.

[4] However, it has been shown that even in a large block of ordinary uranium no chain reaction would take place since too many neutrons would be slowed down by inelastic scattering into the energy region where they are strongly absorbed by U_{238}.

[5] Several people have tried to make a chain reaction possible by mixing the uranium with water, which reduces the energy of the neutrons still further and thereby increases their efficiency again. It seems fairly certain, however, that even then it is impossible to sustain a chan reaction.

[6] In any case, no arrangement containing hydrogen and based on the action of slow neutrons could act as an effective super-bomb, because the reaction would be too slow. The time required to slow down a neutron is about 10^{-5} s and the average time lost before a neutron hits a uranium nucleus is even 10^{-4} s. In the reaction, the number of neutrons would increase exponentially, like $e^{t/\tau}$ where τ would be at least 10^{-4} s. When the temperature reaches several thousand degrees the container of the bomb will break and within 10^{-4} s the uranium would have expanded sufficiently to let the neutrons escape and so to stop the reaction. The energy liberated would, therefore, be only a few times the energy required to break the container, i.e. of the same order of magnitude as with ordinary high explosives.

[7] Bohr has put forward strong arguments for the suggestion that the fission observed with slow neutrons is to be ascribed to the rare isotope U_{235}, and that this isotope has, on the whole, a much greater fission probability than the common isotope U_{238}. Effective methods for the separation of isotopes have been developed recently, of which the method of thermal diffusion is simple enough to permit separation on a fairly large scale.

[8] This permits, in principle, the use of nearly pure U_{235} in such a bomb, a possibility which apparently has not so far been seriously considered. We have discussed this possibility and come to the conclusion that a moderate amount of U_{235} would indeed constitute an extremely efficient explosive.

[9] The behavior of U_{235} under bombardment with fast neutrons is not known experimentally, but from rather simple theoretical arguments it can be concluded that almost every collision produces fission and that neutrons of any energy are effective. Therefore it is not necessary to add hydrogen, and the reaction, depending on the action of fast neutrons, develops with very great rapidity so that a considerable part of the total energy is liberated before the reactions gets stopped on account of the expansion of the material.

[10] The critical radius r_o—that is, the radius of a sphere in which the surplus of neutrons created by the fission is just equal to the loss of neutrons by escape through the surface—is, for a material with a given composition, in a fixed ratio to the mean free path of the neutrons, and this in turn is inversely proportional to the density. It therefore pays to bring the material into the densest possible form, i.e. the metallic state, probably sintered or hammered. If we assume, for U_{235}, no appreciable scattering, and 2.3 neutrons emitted per fission, then the critical radius is found to

Appendix B: The Memorandum: Technical Part 93

be 0.8 times the mean free path. In the metallic state (density 15), and assuming a fission cross-section of 10^{-23} cm^2, the mean free path would be 2.6 cm and r_o would be 2.1 cm, corresponding to a mass of 600 grams. A sphere of metallic U$_{235}$ of a radius greater than r_o would be explosive, and one might think of about 1 Kg as a suitable size for the bomb.

[11] The speed of the reaction is easy to estimate. The neutrons emitted in the fission have velocities of about 10^9 cm/s and they have to travel 2.6 cm before hitting a uranium nucleus. For a sphere well above the critical size the loss through neutron escape would be small, so we may assume that each neutron, after a life of 2.6×10^{-9} s, produces fission, giving birth to two neutrons. In the expression $e^{t/\tau}$ for the increase of neutron density with time, τ would be about 4×10^{-9} s, very much shorter than in the case of chain reaction depending on slow neutrons.

[12] If the reaction proceeds until most of the uranium is used up, temperatures of the order of 10^{10} degrees and pressures of about 10^{13} atmospheres are produced. It is difficult to predict accurately the behavior of matter under these extreme conditions, and the mathematical difficulties of the problem are considerable. By a rough calculation we get the following expression for the energy liberated before the mass expands so much that the reaction is interrupted:

$$E = 0.2 M (r^2/\tau^2)(\sqrt{r/r_o} - 1) \tag{1}$$

(M total mass of uranium; r, radius of sphere; r_o critical radius, τ time required for neutron density to multiply by a factor e). For a sphere of radius 4.2 cm ($r_o = 2.1$ cm), $M = 4700$ grams, $\tau = 4 \times 10^{-9}$ s, we find $E = 4 \times 10^{20}$ ergs, which is about one tenth of the total fission energy. For a radius of about 8 cms ($M = 32$ Kg) the whole fission energy is liberated, according to formula (1). For small radii the efficiency falls off even faster than indicated by formula (1) because τ goes up as r approaches r_o. The energy liberated by a 5 kg bomb would be equivalent to that of several thousand tons of dynamite, while that of a 1 kg bomb, though about 500 times less, would still be formidable.

[13] It is necessary that such a sphere should be made in two (or more) parts which are brought together first when the explosion is wanted. Once assembled, the bomb would explode within a second or less, since one neutron is sufficient to start the reaction and there are several neutrons passing through the bomb in every second, from the cosmic radiation. (Neutrons originating from the action of uranium alpha rays on light-element impurities would be negligible provided the uranium is reasonably pure.) A sphere with a radius of less than about 3 cm could be made up in two hemispheres, which are pulled together by springs and kept separated by a suitable structure which is removed at the desired moment. A larger sphere would have to be composed of more than two parts, if the parts, taken separately, are to be stable.

[14] It is important that the assembling of the parts should be done as rapidly as possible, in order to minimize the chance of a reaction getting started at a moment when the critical conditions have only just been reached. If this happened, the reaction

rate would be much slower and the energy liberation would be considerably reduced; it would, however, always be sufficient to destroy the bomb.

[15] It may be well to emphasize that a sphere only slightly below the critical size is entirely safe and harmless. By experimenting with spheres of gradually increasing size and measuring the number of neutrons emerging from them under a known neutron bombardment, one could accurately determine the critical size, without any danger of a premature explosion.

[16] For the separation of the U_{235}, the method of thermal diffusion, developed by Clusius and others, seems to be the only one which can cope with the large amounts required. A gaseous uranium compound, for example, uranium hexafluoride, is placed between two vertical surfaces which are kept at a different temperature. The light isotope tends to get more concentrated near the hot surface, where it is carried upwards by the convection current. Exchange with the current moving downwards along the cold surface produces a fractionating effect, and after some time a state of equilibrium is reached when the gas near the upper end contains markedly more of the light isotope than near the lower end.

[17] For example, a system of two concentric tubes, of 2 mm separation and 3 cm diameter, 150 cm long, would produce a difference of about 40% in the concentration of the rare isotope between its ends, and about 1 gram per day could be drawn from the upper end without unduly upsetting the equilibrium.

[18] In order to produce large amounts of highly concentrated U_{235}, a great number of these separating units will have to be used, being arranged in parallel as well as in series. For a daily production of 100 grams of U_{235} of 90% purity, we estimate that about 100,000 of these tubes would be required. This seems a large number, but it would undoubtedly be possible to design some kind of a system which would have the same effective area in a more compact and less expensive form.

[19] In addition to the destructive effect of the explosion itself, the whole material of the bomb would be transformed into a highly radioactive state. The energy radiated by these active substances will amount to about 20% of the energy liberated in the explosion, and the radiation would be fatal to living beings even long time after the explosion.

[20] The fission of uranium results in the formation of great number of active bodies with periods between, roughly speaking, a second and a year. The resulting radiation is found to decay in such a way that the intensity is about inversely proportional to the time. Even one day after the explosion the radiation will correspond to a power expenditure of the order of 1000 KW, or to the radiation of a hundred tons of radium.

[21] Any estimates of the effects of this radiation on human beings must be rather uncertain because it is difficult to tell what will happen to the radioactive material after the explosion. Most of it will probably be blown into the air and carried away by the wind. This cloud of radioactive material will kill everybody within a strip estimated to be several miles long. If it rained the danger would be even worse because active material would be carried down to the ground and stick to it, and persons entering the contaminated area would be subjected to dangerous radiations even after days. If 1% of the active material sticks to the debris in the vicinity of the

explosion and if the debris is spread over an area of, say, a square mile, any person entering this area would be in serious danger, even several days after the explosion.

[22] In these estimates, the lethal dose of penetrating radiation was assumed to be 1000 Roentgen; consultation of a medical specialist on X-ray treatment and perhaps further biological research may enable one to fix the danger limit more accurately. The main source of uncertainty is our lack of knowledge as to the behaviour of materials in such a super-explosion, and an expert on high explosives may be able to clarify some of these problems.

[23] Effective protection is hardly possible. Houses would offer protection only at the margins of the danger zone. Deep cellars or tunnels may be comparatively safe from the effects of the radiation, provided air can be supplied from an uncontaminated area (some of the active substances would be noble gases which are not stopped by ordinary filters).

[24] The irradiation is not felt until hours later when it may be too late. Therefore it would be very important to have an organisation which determines the exact extent of the danger area, by means of ionisation measurements, so that people can be warned from entering it.

O. R. Frisch
R. Peierls

The University, Birmingham

Appendix C
Derivation: The Critical Mass

This chapter presents a derivation of the expression for the critical mass of an untamped bomb core. This differs from Peierls' 1939 analysis, but the equivalence of his derivation and the one given here in the parameter range relevant for a practical weapon is developed in Appendix E.

A key step in Frisch and Peierls' analysis was the assumption that the critical radius for a spherical bomb core is approximately 0.8 times the mean free path for fission when the number of neutrons per fission is 2.3 and scattering can be neglected. This Appendix develops a derivation of this claim.

The mean free path is given by

$$\lambda_f = \frac{1}{\sigma_f n}, \tag{C.1}$$

where σ_f is the fission cross section and n the number density of nuclei. This is Eq. (4.1) for the specific case of the fission cross section.

In his 1939 criticality analysis, Peierls used an approach of solving an integral equation for the evolution of the population of neutrons as a function of position and time within an exploding bomb core. Ultimately, one is looking for the minimalist conditions that give a just-growing neutron population; this is the definition of criticality. The integral equation approach gives very accurate results, but is mathematically complex. In this Appendix I analyze the factor of 0.8 via diffusion theory adopted from Serber's *The Los Alamos Primer* and presented in my own *The Physics of the Manhattan Project.* This gives somewhat less accurate but still quite respectable results in a more straightforward way. In particular, this approach will prove familiar to students who have studied some differential equations. For completeness, the essential results of Peierls' approach are summarized at the end of this Appendix, and the equivalence of the diffusion method and Serber's approach in a restricted case is examined in Appendix E.

Many of the results developed in this Appendix will be useful in Appendix D; the two are best read together.

The neutron diffusion equation is a differential equation that describes the space/time behavior of the neutron number density N, that is, the number of neutrons per cubic meter within the bomb core. A derivation of the equation appears in the references cited above. Be sure not to confuse n and N; n is the number density of fissile *nuclei*, while N is the number density of *neutrons*; both play roles in what follows.

A few comments on the limitations of diffusion theory are appropriate here. A diffusion approach is valid if neutron scattering from nuclei is isotropic. Even if scattering is not isotropic, a diffusion approach will still be reasonable if neutrons suffer enough scatterings so as to effectively erase non-isotropic effects. Unfortunately, these conditions are not fulfilled in the case of a uranium (or plutonium) bomb core, where neutron scattering shows a forward-peaked effect. Further, since the mean free path of a fast neutron in ^{235}U is of only about the same order of magnitude as the critical radius itself, one cannot help but question the inherent accuracy of a diffusion model. However, it turns out that with modern numbers and proper recognition of scattering effects, the predictions of diffusion theory compare quite favorably with experimentally-measured critical masses. It is certainly good enough for our purposes.

The analysis here uses spherical coordinates; r is the usual radial distance as measured from the center of the bomb core, and t is time as measured from the moment the nuclear explosion is triggered. The diffusion equation is

$$\frac{\partial N}{\partial t} = \frac{\langle v \rangle}{\lambda_f}(\nu - 1)N + \frac{\lambda_f \langle v \rangle}{3}\left[\frac{1}{r^2}\frac{\partial}{\partial r}\left(r^2 \frac{\partial N}{\partial r}\right)\right]. \tag{C.2}$$

In this expression, $\langle v \rangle$ is the average speed of a fission-liberated neutron [see Eq. (4.5)] and ν the number of neutrons liberated per fission. The left side of the equation expresses the overall rate of change of neutron density as the sum of the two contributions on the right side. The first term on the right side is the growth rate in the neutron density due to fissions; the factor of $(\nu - 1)$ appears because one neutron is always consumed in causing a fission. The second term on the right accounts for the rate of neutron loss by their flying out of a volume of space being considered. This version of the diffusion equation assumes no scattering, per Frisch and Peierls; if scattering is involved, the factor of λ_f in the last term has to be replaced with the so-called transport or total mean free path; see Appendix E. For all of what follows in this Appendix, we assume no scattering.

Appendix C: Derivation: The Critical Mass

Our problem is to solve for N and then apply an appropriate neutron-flux boundary condition at the edge of the core. Application of the condition results in an equation of constraint for the critical radius.

This differential equation involves two variables, r and t. The usual strategy for solving such an equation is to posit a trial solution of a separated form, that is, where the effects of r and t are embedded in separate functions that depend upon only one variable each, with the overall solution being the product of the functions: $N(r, t) = N_r(r) N_t(t)$. If we can find a solution to the differential equation with this assumption, mathematicians assure us that it must be correct as all solutions to a given differential equation must be equivalent.

The convenience of the separation assumption is that, so far as derivatives with respect to r are concerned, $N_t(t)$ acts like a constant, and likewise for $N_r(r)$ so far as derivatives with respect to t are concerned. Upon substituting the trial solution, accounting for these "flow-through" behaviors, and then dividing through by $N_r(r) N_t(t)$, Eq. (C.2) becomes

$$\frac{1}{N_t}\left(\frac{\partial N_t}{\partial t}\right) = \left(\frac{\nu - 1}{\tau_o}\right) + \frac{D}{N_r}\left[\frac{1}{r^2}\frac{\partial}{\partial r}\left(r^2 \frac{\partial N_r}{\partial r}\right)\right], \quad (C.3)$$

where τ_o is the average time that a neutron will travel before causing a fission [see Eq. (4.12)]:

$$\tau_o = \frac{\lambda_f}{\langle v \rangle}, \quad (C.4)$$

and where D is the so-called diffusion coefficient,

$$D = \frac{\lambda_f \langle v \rangle}{3} = \frac{\lambda_f^2}{3\tau_o}, \quad (C.5)$$

where in the last step we used Eq. (C.4) to eliminate $\langle v \rangle$. D has units of m^2/s.

Equation (C.3) is separated: The left side depends only on t and the right side depends only on r. The additive constant in large round brackets on the right side could be placed on either side, but custom is to leave it on the radial side to simplify algebra about to appear.

Since r and t are independent variables, such a separated equation can only be satisfied if both sides are equal to a constant, the so-called "separation constant". Both terms in Eq. (C.3) have units of inverse time, so the constant must have the same units. The natural timescale in this problem is τ_o, so it is customary to write the separation constant as α/τ_o, where α is a dimensionless constant yet to be determined. Hence we have

$$\frac{1}{N_t}\left(\frac{\partial N_t}{\partial t}\right) = \left(\frac{\nu - 1}{\tau_o}\right) + \frac{D}{N_r}\left[\frac{1}{r^2}\frac{\partial}{\partial r}\left(r^2 \frac{\partial N_r}{\partial r}\right)\right] = \frac{\alpha}{\tau_o}. \quad (C.6)$$

100   Appendix C: Derivation: The Critical Mass

Consider first the time-dependent part:

$$\frac{1}{N_t}\left(\frac{\partial N_t}{\partial t}\right) = \frac{\alpha}{\tau_o}. \tag{C.7}$$

The solution to this is

$$N_t(t) = N_o e^{(\alpha/\tau_o)t}, \tag{C.8}$$

where N_o represents the neutron density at the center of the core at $t = 0$. N_o would be set by whatever device is used to initiate the chain reaction; we will not need to specify a value for N_o.

Equation (C.8) shows that the time-growth or decay of the neutron density is exponential: Growth if $\alpha > 0$ ("supercritical"), decay if $\alpha < 0$ ("subcritical"). If $\alpha = 0$, one has minimal "threshold criticality," which will be assumed shortly. While our main concern for the present is with the *spatial* behavior of N, the time-evolution of $N(r, t)$ will prove to be particularly important in developing an expression for the energy yield of a bomb in Appendix D.

The radial equation for the neutron flux is

$$\left(\frac{\nu-1}{\tau_o}\right) + \frac{D}{N_r}\left[\frac{1}{r^2}\frac{\partial}{\partial r}\left(r^2\frac{\partial N_r}{\partial r}\right)\right] = \frac{\alpha}{\tau_o}. \tag{C.9}$$

The first and last terms in this expression can be combined by bringing the right-side term over to the left and combining it with the first term on the left; they have the same denominator; this is why the separation constant was defined as α/τ_o. On doing so and dividing through by D of Eq. (C.5), we find

$$\frac{1}{d^2} + \frac{1}{N_r}\left[\frac{1}{r^2}\frac{\partial}{\partial r}\left(r^2\frac{\partial N_r}{\partial r}\right)\right] = 0, \tag{C.10}$$

where d, which has units of length, is defined as

$$d = \frac{\lambda_f}{\sqrt{3(\nu-1-\alpha)}}. \tag{C.11}$$

The point of introducing d is that it allows us to from a dimensionless radial variable:

$$x = \frac{r}{d}. \tag{C.12}$$

On incorporating this into Eq. (C.10) (try it!), you will find that d cancels. Then move the factor of 1 that arises from $1/d^2$ over to the right side. The result is the simpler form

Appendix C: Derivation: The Critical Mass

$$\frac{1}{N_r}\left[\frac{1}{x^2}\frac{\partial}{\partial x}\left(x^2\frac{\partial N_r}{\partial x}\right)\right] = -1. \tag{C.13}$$

The idea here is to now think of N as being a function of x, $N(x)$. The general solution of this differential equation is (verify!)

$$N_r(r) = A\left(\frac{\sin x}{x}\right) + B\left(\frac{\cos x}{x}\right), \tag{C.14}$$

where A and B are constants of integration. There are two terms in the solution because Eq. (C.13) is a second-order differential equation. However, both terms need not apply in any given physical situation, and in this case we drop the $(\cos x / x)$ part of the solution because it would diverge at $x = 0$, that is, at $r = 0$. Not all solutions to a differential equation need be physically reasonable! Also, the remaining constant of integration A will soon cancel out, so we need not worry about specifying it.

To determine a critical radius, we need a boundary condition to apply to Eq. (C.14). For a reason that will be explained shortly, the symbol R_{co} is adopted for this radius. As explained in the references cited above, an accounting of the neutron flux at the edge of the core shows that this boundary condition takes the form

$$N_r(R_{co}) = -\frac{2\lambda_f}{3}\left(\frac{\partial N_r}{\partial r}\right)_{R_{co}} = -\frac{2\lambda_f}{3d}\left(\frac{\partial N_r}{\partial x}\right)_{R_{co}}, \tag{C.15}$$

where in the last step we converted to a derivative with respect to x.

On applying this condition to Eq. (C.14) (remember to drop the cosine term), the constant of integration A cancels, and what remains is a transcendental equation for the critical radius:

$$x\cot(x) + kx - 1 = 0, \tag{C.16}$$

where

$$k = \frac{3d}{2\lambda_f} = \frac{1}{2}\sqrt{\frac{3}{(\nu - 1 - \alpha)}}. \tag{C.17}$$

With fixed values for the density and nuclear constants for some fissile material, Eqs. (C.16) and (C.17) contain two variables: The core radius r (through x) and the exponential factor α (through k). These two equations can be solved in two different ways. For both approaches, assume, as did Frisch and Peierls, that we are working with material of "normal" density, which we designate as ρ_o. For the first approach, start by looking back to Eq. (C.8). If $\alpha = 0$, that is, if there is no growth or decay in the neutron population, we have threshold criticality. To determine the so-called threshold bare critical radius R_{co}, set $\alpha = 0$ in Eqs. (C.11) and (C.17), set the density to ρ_o to determine n and λ_f, solve Eq. (C.16) for x, and then get r ($= R_{co}$) from Eq. (C.12). Now you know why the subscript co was chosen for the critical radius corresponding to normal density. It is this radius that one usually sees referred to as *the* critical radius, although we will see below that the term is not uniquely defined. In Appendix D, the density of the core will change as it expands due to energy released during the explosion.

The other approach to treating the solution is to start with a given core radius $R_0 > R_{co}$, set $x = R_0/d$, and instead solve for α. In this case one has a "supercritical" core as described above. However, our concern here is with the threshold core condition; supercriticality is treated in Appendix D. For completeness, an approximate expression for α in the case of a core whose initial radius is greater than R_{co} is developed at the end of this Appendix.

Before proceeding to recover the factor of 0.8, an important aspect of criticality needs to be pointed out. The quantity x to be solved for in Eq. (C.16) is proportional to r/λ_f; look at Eqs. (C.11) and (C.12). But λ_f is inversely proportional to the density through n [Eq. (C.1)], so the combination r/λ_f is proportional to ρr, that is, the solution of (C.16) demands a unique value of ρr for a given combination of σ_f and ν. *This means that if R_{co} is the bare threshold critical radius for material of normal density ρ_o, then any combination of r and ρ such that $\rho r = \rho_o R_{co}$ will also be threshold critical, and any combination such that $\rho r > \rho_o R_{co}$ will be supercritical.* This explains the remark above that critical radius is not uniquely defined. This "ρr" behavior will be very important in Appendix D.

Now to the factor of 0.8. Knowing that this is what we are after, it is helpful to define a new variable, η, to be the number by which λ_f has to be multiplied to give r:

$$r = \eta \lambda_f. \tag{C.18}$$

Replacing r in the definition of x, the criticality constraint of Eq. (C.16) can be expressed as

$$\eta \gamma \cot(\eta \gamma) + \frac{3}{2}\eta - 1 = 0, \tag{C.19}$$

where, with $\alpha = 0$ for threshold criticality in the definition of k (that is, Eq. (C.18) becomes $R_{co} = \eta \lambda_f$),

$$\gamma = \sqrt{3(\nu - 1)}. \tag{C.20}$$

Once ν is specified, Eq. (C.19) can be solved for η with any spreadsheet or root-finding routine.

Figure C.1 shows $\eta = R_{co}/\lambda_f$ versus ν. For $\nu = 2.3$, $R_{co}/\lambda_f \sim 0.89$, somewhat higher than Frisch and Peierls's adopted value of 0.8 but close enough to make the point. Notice that the critical radius decreases as ν increases, which makes intuitive sense. Also, as $\nu \to 1$, $R_{co} \to \infty$. This too makes sense: If on average only a little more than one neutron is emitted per fission, you will need a large core to get any appreciable growth in the neutron population; losing only a few neutrons to the outside world would suppress the reaction.

Appendix C: Derivation: The Critical Mass

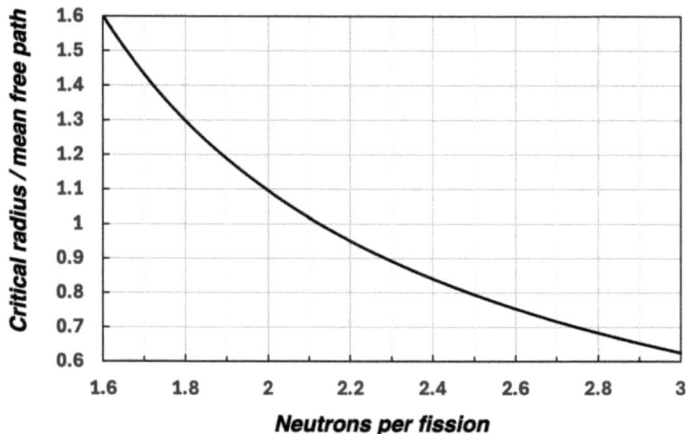

Fig. C.1 Solution of Eq. (C.19) for R_{co}/λ_f versus number of neutrons per fission ν. No scattering

We now have a look at the results of Peierls' more sophisticated calculation; see also Appendix E.

Peierls developed his analysis in terms of a dimensionless parameter ξ which, in its most general form, combines ν and the cross sections for fission and scattering. If scattering is disregarded, however, ξ has the particularly simple form:

$$\xi^2 = \frac{\nu - 1}{\nu}. \tag{C.21}$$

If ν is large (small critical radius), then $\xi \to 1$, whereas if $\nu \to 1$ (large critical radius), $\xi \to 0$. His analysis showed that the threshold critical radius R_{co} (that is, again assuming normal density) is given by, to second order in ξ,

$$\frac{1}{R_{co}} \sim \frac{\nu}{\lambda_f} \times \begin{cases} 0.552\,\xi + 0.216\,\xi^2 & (\nu \sim 1;\ \xi \to 0) \\ 0.78 - 1.02(1 - \xi) & (\nu \gg 1;\ \xi \to 1). \end{cases} \tag{C.22}$$

For Frisch and Peierls' choice of $\nu = 2.3$, $\xi \sim 0.752$. Equation (C.22) then becomes

$$\frac{1}{R_{co}} \sim \frac{2.3}{\lambda_f} \times \begin{cases} 0.537 & (\xi \to 0) \\ 0.527 & (\xi \to 1). \end{cases} \tag{C.23}$$

Taking the average value of the extremes as 0.532 gives $R_{co}/\lambda_f \sim 0.817$, which they evidently rounded to 0.8. For their density and $\sigma_f = 10$ b, $\lambda_f = 2.60$ cm, and the critical radius evaluates as 2.1 cm, as they claimed.

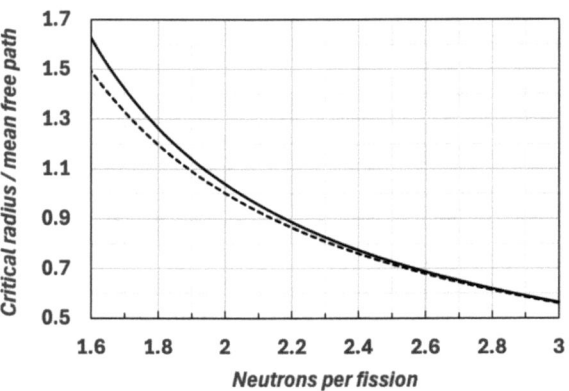

Fig. C.2 R_c/λ_f versus ν for Peierls' criticality analysis with no scattering. The solid curve is valid for $\xi \to 1$, that is, for $\nu \gg 1$, and the dashed curve for $\xi \to 0$ ($\nu \to 1$)

Figure C.2 shows R_{co}/λ_f versus ν for this formulation; clearly, the two curves corresponding to $\nu \sim 1$ and $\nu \gg 1$ do not differ significantly over a sensible range of ν.

A question: This figure indicates that the critical radius will be *greater* for the expression corresponding to $\xi \to 1$ than for that corresponding to $\xi \to 0$, whereas one would expect the opposite. What do you suppose might be going on here?

Finally, we return to establishing an approximate estimate for the separation constant α of Eqs. (C.6) and (C.7) in the case of a core whose initial radius is greater than the critical radius R_{co}.

As a simplified boundary condition for this purpose, assume that $N_r(r_{edge}) = 0$, that is, that the neutron density falls to zero at the edge of the core. This is a more restrictive condition than the true boundary condition and will lead to a larger critical radius, but the point here is to make a rough estimate for α. Now suppose that we have a core of radius $r = R_0$. In this case, Eq. (C.14) indicates that we must have $\sin(x) = 0$. This demands $x = \pi$, or $d = R_0/x = R_0/\pi$. Back-substitute this expression for d into Eq. (C.11) and solve for α:

$$\alpha = (\nu - 1) - \frac{\lambda_f^2 \pi^2}{3 R_0^2}. \tag{C.24}$$

If the core radius R_0 is very small, then we will have $\alpha < 0$ because of the second term in this expression, and the neutron population will decline in time in Eq. (C.8). Threshold criticality corresponds to $\alpha = 0$, in which case R_0 is labeled as R_{co}:

$$R_{co}^2 = \frac{\lambda_f^2 \pi^2}{3(\nu - 1)}. \tag{C.25}$$

This is a very approximate expression for R_{co}; it differs from the earlier analysis described following Eq. (C.17) because of the different boundary condition. To get an approximate expression for α in terms of the critical radius R_{co} and a core of and radius R_0 (presumed to be greater than R_{co}), use this result to eliminate $\lambda_f^2 \pi^2$ in (C.24) to give

$$\alpha \sim (\nu - 1)\left[1 - \left(\frac{R_{co}}{R_0}\right)^2\right]. \tag{C.26}$$

It is important to emphasize that Eq. (C.26) will be the value of α at the moment when the nuclear explosion is initiated. As discussed in Appendix D, as the core expands and approaches its criticality-shutdown radius, α will decrease to zero.

Important: The analysis here is equivalent to assuming that the number of neutrons liberated per fission is just slightly greater than one, in which case the core will be just marginally critical and will have to be very large; look at the denominator in Eq. (C.25) for $\nu \sim 1$. This is equivalent to Peierls' analysis for the case of $\xi \to 0$; look at Eq. (C.21). Hence, Peierls' equation (C.22) and the criticality constraint from diffusion analysis, Eq. (C.16), should give the same result for the critical radius even if scattering and fission both occur. A proof of this appears in Appendix E. The algebra is tedious and somewhat convoluted, but it does work out. Unfortunately, the same sort of analysis cannot be used for the case of a small critical radius: The diffusion theory breaks down in that case because the core will be so small that isotropization of neutron paths by scattering cannot reasonably be expected to occur.

Appendix D
Derivation: Weapon Yield

This offers a *possible* reconstruction of Frisch and Peierls' expression for the energy yield of a nuclear weapon that appears in the quantitative part of the memorandum. The emphasis on "possible" here reflects the fact that they gave no indication of how they arrived at their result.

This Appendix offers a derivation of the only expression appearing in the technical part of the FP memorandum, a formula for the energy E released in the explosion of a fission bomb. In the notation of Appendix C, this is

$$E = 0.2 \, M_{core} \left(\frac{R_0^2}{\tau^2}\right) \left[\sqrt{\frac{R_0}{R_{co}}} - 1\right]. \quad \text{(D.1)}$$

Here M_{core} is the total mass of the core, R_0 its initial radius, R_{co} the critical radius for the fissile material concerned (uncompressed ^{235}U), and τ the exponential growth time of Eq. (4.13).

Since Frisch and Peierls gave no indication as to how they arrived at this expression, it must be emphasized that the following derivation should be taken in the sense of suggesting what they *might* have done. Along the way, various simplifying approximations will be made; the point here is to reproduce the overall form of Eq. (D.1). During the war, various weapon yield/efficiency formulae were developed; as described by Lestone et al. (2021), these all have the same form as the

Frisch–Peierls expression, although they differ in the leading numerical factor. This concurrence is presumably reflective of the same physical concepts underlying the various derivations. The derivation presented here is adapted from Pearson and Reed (2024).

The derivation involves four steps: (i) As a bomb core explodes, it rapidly heats and expands. This causes the density to drop. Recall from Appendix C that the state of criticality (i.e., an ongoing chain reaction) is dictated by the product of instantaneous density and radius, ρr. Inevitably, the core expands to a radius where criticality no longer holds. The first part of the derivation develops an expression for the criticality-shutdown radius R_{shut} in terms of R_0 and R_{co}. (ii) The neutron population model of Appendix C is used to get an expression for the total energy liberated by fissions up to a general time t after the initiation of the explosion. (iii) An idealized model of the rate of expansion of the core is posited. This is used to establish an estimate of the elapsed time required for the core to reach shutdown radius, t_{shut}. (iv) Finally, t_{shut} is used in the energy expression of step (ii) to assess the energy liberated to that time.

(i) The criticality shutdown radius

An expression for the core radius at criticality shutdown can be established from the ρr argument summarized above. Here it is assumed that the core does not lose any appreciable amount of mass during the explosion. This may sound strange, but is not unrealistic for the type of weapon Frisch and Peierls modeled: The Hiroshima uranium bomb was only about 1–2% efficient. Very little of the expensive fissile material actually fissioned; the remainder ended up as fallout. As in Appendix C, for material of normal density ρ_o and threshold critical mass R_{co}, the ρr criticality product will be $\rho_o R_{co}$.

Now suppose that we have a core of initial radius $R_0 > R_{co}$. The mass will be $M_{core} = 4\pi R_0^3 \rho_o / 3$. At any general expanded radius R, the density ρ must then be such that $4\pi R^3 \rho / 3 = M_{core}$, or, $\rho = R_0^3 \rho_o / R^3$. The criticality product ρR is then $\rho R = R_0^3 \rho_o / R^2$. When R has grown to the point that this equals $\rho_o R_{co}$, we must be at the shutdown radius:

$$\frac{R_0^3 \rho_o}{R_{shut}^2} = \rho_o R_{co} \Rightarrow R_{shut} = R_0 \sqrt{\frac{R_0}{R_{co}}}. \tag{D.2}$$

This may look like a somewhat odd way of expressing the result, but it will prove helpful later in recovering the form of Eq. (D.1).

Throughout this Appendix, there will be various versions of the core radius to keep in mind. A summary: R_{co} is the threshold critical radius for the material involved, whose density is ρ_o. R_0 is the initial radius of the core, $R_0 > R_{co}$. R_{shut} is the expanded radius at which criticality shuts down. R without a subscript means a general outer radius of the core at some time; think of it as $R(t)$. As the explosion proceeds, $R(t)$

Appendix D: Derivation: Weapon Yield

evolves from R_0 to R_{shut}. The symbol r is used for a general radius *within* the core at some time.

Cores typically expand by only a small amount before losing criticality. For example, if you have a core of two critical *masses*, $R_0 = 2^{1/3} R_{co}$, and so $R_{shut} = 2^{1/6} R_0 = 1.12 R_0$: There is only a 12% expansion in radius until shutdown occurs. To nuclear weapons engineers, shutdown is known as "second criticality"; first criticality is the moment when criticality first occurs when the bomb is triggered.

A point regarding the exponential timescale τ is worth making here. This timescale is inversely proportional to the density of the core material, and hence will grow as the core expands and the density drops. In the following algebra we will treat τ as being constant, another level of approximation. A rough estimate of the growth of τ from its initial value of τ_o is treated in Chap. 4, but the essential point here is that there is no strictly rigorous analytic treatment of bomb yield available; essentially all of the variables are time-dependent in non-linear (mostly exponential) ways. At Los Alamos, manual calculations, electromechanical calculators (essentially glorified adding machines), and early computers were used to carry out lengthy numerical simulations of exploding cores. Again, the point here is to see how Frisch and Peierls *might* have arrived at their expression.

(ii) Neutron population and energy liberation

Begin by combining Eqs. (C.8) and (C.14) with $\tau = \tau_o/\alpha$ to give the overall space and time evolution of the neutron density $N(r, t)$, again dropping the cosine term in (C.14):

$$N(r, t) = N_o e^{(t/\tau)} \left(\frac{\sin x}{x} \right), \tag{D.3}$$

where we have written the timescale τ in the exponential per the discussion in Appendix C and where x is given by Eqs. (C.11) and (C.12). The important one of the latter here is (C.12):

$$x = \frac{r}{d}. \tag{D.4}$$

At this point an approximation is invoked that lets us write Eq. (D.3) in a form that will be easier to deal with as we go forward. This is to employ the simpler boundary condition described following Eq. (C.23): That the neutron density falls to zero at the edge of the core, $N_r(R_{core}) = 0$. As described therein, Eq. (C.14) or (D.3) then indicates that we must have $\sin(x) = 0$ at the surface of the core, or $x = r/d = \pi$. With this we can set $d = R/\pi$, and hence $x = \pi r/R$. Reminder: R now represents the outer radius of the core, which is strictly a function of time. This said, there will be occasions where we will approximate R as R_0, but the intent is to try to retain as much generality as possible. Equation (D.3) then becomes

$$N(r,t) = N_o e^{(t/\tau)} \left(\frac{R}{r\pi}\right) \sin\left(\frac{\pi r}{R}\right) = \left(\frac{N_o R}{\pi}\right) \frac{e^{(t/\tau)}}{r} \sin\left(\frac{\pi r}{R}\right). \quad (D.5)$$

The factors of R and π in the first round brackets could be absorbed into N_o, but they will be left as is for sake of explicitness. As before, we will not need to know N_o explicitly.

To determine the total number of neutrons in the core at any time, integrate Eq. (D.5) from $r = 0$ to $r = R$. As usual with spherical problems, divide the sphere into shells of thickness dr; the volume of a shell at radius r is $4\pi r^2 dr$, and the number of neutrons in that shell will be $4\pi N(r,t)r^2 dr$. Using the symbol $\mathcal{N}(t)$ to represent the total number of neutrons as a function of time, we get

$$\mathcal{N}(t) = 4\pi \int_0^R N(r,t) r^2 dr = 4 N_o R\, e^{(t/\tau)} \int_0^R r \sin\left(\frac{\pi r}{R}\right) dr.$$

The integral here evaluates to R^2/π, giving

$$\mathcal{N}(t) = \frac{4 N_o}{\pi} R^3 e^{(t/\tau)}. \quad (D.6)$$

We can use this result to get an expression for the total energy liberated by fissions up to time t. Each fission liberates ν neutrons, but consumes one neutron in the process. Hence there will be a net gain of $(\nu - 1)$ neutrons per fission. The number of fissions that must have happened is then $\mathcal{N}(t)/(\nu - 1)$. If each fission liberates energy ϵ, we must have

$$E_{fiss}(t) = \frac{4 \epsilon N_o}{\pi (\nu - 1)} R^3\, e^{(t/\tau)}. \quad (D.7)$$

Now, what will be useful subsequently is an expression for the *energy density* $U(t)$, that is, the total energy liberated to time t divided by the volume of the core, $4\pi R^3 / 3$:

$$U(t) = \frac{3 \epsilon N_o}{\pi^2 (\nu - 1)} e^{(t/\tau)}. \quad (D.8)$$

The reason for introducing the energy density is that it gives us a way of determining the pressure within the exploding core, which will be useful in the following subsection where we consider the dynamics of the explosion.

Results from thermodynamics and statistical mechanics show that for a gas of material particles or photons, pressure is related to energy density via $P = \gamma U$. The value of the constant γ depends on whether ordinary gas pressure ($\gamma = 2/3$) or radiation pressure in the case of photons ($\gamma = 1/3$) is dominant. In an exploding bomb core, both will be present: The fission products act like a high-speed gas, but as they are slowed by collisions with and causing ionization of other nuclei, they emit copious amounts of electromagnetic radiation. Instead of trying to specify γ,

Appendix D: Derivation: Weapon Yield

we leave it as a factor in the equations to see how it flows through the algebra. The pressure will then behave as

$$P(t) = \left(\frac{3\epsilon N_o \gamma}{\pi^2 (\nu - 1)}\right) e^{(t/\tau)} = P_o e^{(t/\tau)}, \tag{D.9}$$

where P_o represents the factors in the large round brackets.

What is needed now is an estimate for the amount of time that passes until criticality shutdown, t_{shut}; the yield of the bomb will then be $E_{fiss}(t_{shut})$. This requires a model for the expansion of the core, which is the subject of the following subsection.

(iii) Core expansion

At this point, the discussion moves to consider the dynamics of the expanding core. Without detailed knowledge of the conditions inside an exploding bomb core, we have to adopt some model of how it expands. There are various possibilities. A simple one that comes to mind is to suggest that all points within the core expand at the same speed, but this runs into the question of how to treat a point at the center of the core.

The assumption made here—and it is an operating assumption—is that the expansion is "homologous." In a homologous expansion, the speed as a function of radius behaves as

$$\frac{dr}{dt} = cr. \tag{D.10}$$

The scale factor c, which must have units of inverse time, may be a function of time, $c(t)$. Such an expansion is similar to that of Hubble's law in cosmology, where the Hubble "constant" is a function of time. Presumably one would have to require $c(t) \to 0$ as $t \to 0$ as nothing is moving at the beginning, but this leaves an almost infinite number of possibilities. Fortunately, however, it turns out that we will not have to get this function explicitly in order to derive the desired yield formula.

The reason for adopting such an expansion model is that it leads to the density of the core material always being *uniform* throughout, that is, the same everywhere at a given time, although it steadily decreases as the core expands. To see this, imagine a shell at some initial radius r_o at time $t = 0$. What will be its radius at time t? Equation (D.10) can be separated and integrated:

$$\int_{r_o}^{r(t)} \frac{dr}{r} = \int_0^t c(t) dt,$$

which gives

$$r(t) = r_o e^{\int_0^t c(t) dt} = r_o e^C, \tag{D.11}$$

where the integral in the exponential is compacted as the symbol C; do not confuse this with the function $c(t)$ itself.

To understand the density claim above, imagine a shell of material of initial inner radius r_o and thickness Δr. The initial volume of the shell will be

$$V(t=0) = \frac{4}{3}\pi \left[(r_o + \Delta r)^3 - r_o^3\right].$$

As the shell expands, the inner and outer edges move according as Eq. (D.11). At time t,

$$V(t) = \frac{4}{3}\pi \left[(r_o + \Delta r)^3 - r_o^3\right] e^{3C}.$$

Be sure to understand how the factor of e^{3C} arises.

Here is the key point: With a homologous expansion, greater radii move outward faster than lesser radii. This means that material that was inside r_o cannot migrate into the shell being considered, nor can material within a shell migrate into a surrounding one; *such an expansion will preserve the mass of the shell*. If the density at $t = 0$ is ρ_o and that at time t is $\rho(t)$, we must have $\rho_o V(t=0) = \rho(t) V(t)$ and hence

$$\rho(t) = \rho_o e^{-3C}. \tag{D.12}$$

Notice that r_o does not appear in this result; it must apply for a shell *anywhere* within the sphere: The density is uniform at all times but decreases exponentially in time. We will use this result in what follows.

We now establish an expresion for the kinetic energy of the expanding core at any moment. We can build this up from the usual $mv^2/2$ result from basic physics by again dividing the core up into shells of radius r and thickness dr. From the density discussion above, the mass of a shell of volume $4\pi r^2 dr$ at any moment will be $4\pi r^2 \rho_o e^{-3C} dr$. From the homology model, the speed of the shell will be $v = c(t)r$, so $K_{shell} = \frac{1}{2}(4\pi r^2 \rho_o e^{-3C})[c(t)]^2 dr$. Again writing the radius of the core as R at any moment, integrating over the entire core gives

$$KE_{core} = 2\pi \rho_o e^{-3C} [c(t)]^2 \int_0^R r^4 dr = \frac{2}{5}\pi \rho_o e^{-3C} [c(t)]^2 R^5. \tag{D.13}$$

It is legitimate to factor the time-dependent C and $c(t)$ terms out of the integral because we are imagining a fixed time and integrating over r.

Now, recall that we assume that the mass of the core stays constant during the explosion. We can then write

$$M_{core} = \frac{4}{3}\pi (\rho_o e^{-3C}) R^3. \tag{D.14}$$

Appendix D: Derivation: Weapon Yield

Solve this expression for $\rho_o e^{-3C}$ and back substitute into Eq. (D.13) to write the kinetic energy as

$$KE_{core} = \frac{3}{10} M_{core} [c(t)]^2 R^2. \qquad (D.15)$$

From the homology model we can write, for the edge of the core,

$$c(t) = \frac{1}{R}\left(\frac{dR}{dt}\right) = \frac{v_R}{R}, \qquad (D.16)$$

where $v_R = dR/dt$ is introduced temporarily as the expansion speed of the surface of the core at any moment. Using this to eliminate $c(t)$ in Eq. (D.15) gives

$$KE_{core} = \frac{3}{10} M_{core} v_R^2. \qquad (D.17)$$

This expression is similar to the familiar $mv^2/2$ expression, modified to account for the homologous expansion model. We will need this result shortly.

The expansion of the bomb core must be caused by some force. This can only be the gas/radiation pressure of Eq. (D.9). To study the dynamics of the expansion, we appeal to the classical work-energy theorem in its thermodynamic formulation $W = P(t)dV$, and equate the work W done by the pressure in changing the core volume by dV over time dt to the change in the core's kinetic energy over that time:

$$P(t)\frac{dV}{dt} = \frac{d(KE_{core})}{dt}. \qquad (D.18)$$

For dV/dt we can write

$$\frac{dV}{dt} = \frac{d}{dt}\left(\frac{4}{3}\pi R^3\right) = 4\pi R^2 \left(\frac{dR}{dt}\right). \qquad (D.19)$$

For the rate of change of the kinetic energy, use Eq. (D.17):

$$\frac{d(KE_{core})}{dt} = \frac{3}{5} M_{core} v_R \left(\frac{dv_R}{dt}\right) = \frac{3}{5} M_{core} \left(\frac{dR}{dt}\right)\left(\frac{dv_R}{dt}\right), \qquad (D.20)$$

where in the last step we reverted back to dR/dt notation: $v_R = dR/dt$.

Gather Eqs. (D.19) and (D.20) into (D.18), and use (D.9) for $P(t)$. This gives

$$P_o e^{(t/\tau)} 4\pi R^2 \left(\frac{dR}{dt}\right) = \frac{3}{5} M_{core} \left(\frac{dR}{dt}\right)\left(\frac{dv_R}{dt}\right).$$

The factors of (dR/dt) cancel, leaving

$$\left(\frac{dv_R}{dt}\right) = \left(\frac{20\pi P_o}{3M_{core}}\right) R^2 \, e^{(t/\tau)}. \tag{D.21}$$

This expresses the acceleration of the outer surface of the core.

At this point we make another simplifying approximation. R here will be a function of time, but we cannot know it unless we were to know the homology function $c(t)$ explicitly. However, we know from the argument in subsection (i) above that the core will not expand too much before criticality shutdown. So, this factor gets approximated as the initial radius of the core, R_0. In what follows we *will* worry about the expansion of the core to know when criticality shuts down, so this is admittedly a not an entirely consistent approach, but is one that Frisch and Peierls apparently invoked. The resulting factor of $R^2 \to R_0^2$ is then brought within the brackets in Eq. (D.21) to cast it as

$$\left(\frac{dv_R}{dt}\right) = \left(\frac{20\pi P_o R_0^2}{3M_{core}}\right) e^{(t/\tau)}. \tag{D.22}$$

This expression can be separated and integrated from $t = 0$ (at which time $v_R = 0$) to some later general time t:

$$\int_0^{v_R(t)} dv_R = \left(\frac{20\pi P_o R_0^2}{3M_{core}}\right) \int_0^t e^{(t/\tau)} dt.$$

The left side will give $v_R(t)$, which we can write as dR/dt. On the right side, the integral of the exponential will yield $\tau(e^{t/\tau} - 1)$. The factor of -1 is dropped on the rationale that likely many neutron generations will occur by the time of criticality shutdown, that is, that we will have $e^{t_{shut}/\tau} \gg 1$. This gives

$$\frac{dR}{dt} = \left(\frac{20\pi P_o R_0^2 \tau}{3M_{core}}\right) e^{(t/\tau)}. \tag{D.23}$$

Separate and integrate again, from $t = 0$ (when $R = R_0$) to $t = t_{shut}$ when $R = R_{shut}$:

$$\int_{R_0}^{R_{shut}} dR = \left(\frac{20\pi P_o R_0^2 \tau}{3M_{core}}\right) \int_0^{t_{shut}} e^{(t/\tau)} dt.$$

Again dropping the lower limit on the right side gives

$$(R_{shut} - R_0) = \left(\frac{20\pi P_o R_0^2 \tau^2}{3M_{core}}\right) e^{(t_{shut}/\tau)},$$

Appendix D: Derivation: Weapon Yield

that is,

$$e^{(t_{shut}/\tau)} = \left(\frac{3 M_{core}}{20 \pi P_o R_0 \tau^2}\right)(R_{shut} - R_0),$$

or, on invoking P_o from Eq. (D.9),

$$e^{(t_{shut}/\tau)} = \left(\frac{\pi M_{core}(\nu - 1)}{20 N_o \epsilon \gamma R_0^2 \tau^2}\right)(R_{shut} - R_0). \tag{D.24}$$

(iv) Energy liberated to time t_{shut}

We are almost there. Back-substitute Eq. (D.24) into the energy-liberation formula, Eq. (D.7), to determine the yield to t_{shut}. In doing so, approximate the factor of R^3 in (D.7) as R_0^3:

$$E_{fiss}(t_{shut}) = \frac{4 \epsilon N_o R_0^3}{\pi(\nu - 1)}\left(\frac{\pi M_{core}(\nu - 1)}{20 N_o \epsilon \gamma R_0^2 \tau^2}\right)(R_{shut} - R_0)$$

$$\tag{D.25}$$

$$= \frac{M_{core} R_0}{5 \gamma \tau^2}(R_{shut} - R_0).$$

Notice that N_o, ϵ, and $(\nu - 1)$ canceled here; this will be commented on below. The final step is to use R_{shut} from Eq. (D.2), and draw a factor of R_0 outside the brackets:

$$E_{fiss}(t_{shut}) = \frac{M_{core} R_0^2}{5 \gamma \tau^2}\left(\sqrt{\frac{R_0}{R_{co}}} - 1\right). \tag{D.26}$$

This result is of exactly the form of the Frisch–Peierls formula, Eq. (D.1), with the exception of the presence of the pressure-energy factor γ in the present result.

If Frisch and Peierls followed a development like this, the fact that they set $\gamma \sim 1$ should not be too great a cause for alarm; they were looking for an approximate formula that captured the dependence of the yield on the various parameters involved. An overestimate of γ would drive down the yield estimate, so such an approximation would err on the side of conservatism.

A perhaps surprising aspect of this result is that it is completely independent of the energy released per fission, ϵ. A larger value of ϵ means more energy per fission, but from Eq. (D.24) a smaller time-to-shutdown because ϵ appears in the denominator. The two effects cancel when computing the overall energy release. In two Universes that are otherwise identical but for the value of ϵ, identically-designed bombs would have the same yield, but the bomb in the larger-ϵ Universe would create a more devastating shock wave on account of its briefer explosion time. Also, while it looks like the result is independent of the number of neutrons per fission, this is "hiding" in the timescale τ.

Finally, is it reasonable for us to have assumed that enough fission generations will elapse so that we can take $e^{(t_{shut}/\tau)} \gg 1$? Without getting into the weeds of particular values of critical masses, we can make a quick numerical estimate by asking how long it would take to fission one kilogram of uranium nuclei if on average two neutrons are emitted per fission; this is about the amount of material that underwent fission over Hiroshima. One kilogram of ^{235}U contains about $\Omega \sim 2.56 \times 10^{24}$ nuclei. With $\nu = 2$ neutrons liberated per fission, the number of generations G necessary to fission the entire kilogram would be $\nu^G = \Omega$. Solving for G gives $G = \ln(\Omega)/\ln(\nu) \sim 81$. e^{81} is certainly much greater than 1! 80 generations may sound like a lot, but with each occupying $\tau \sim 10^{-8}$ s, one has an explosion time of on the order of only a microsecond. Nuclear explosions are incomprehensibly brief and powerful; they are literally orders of magnitude faster than the blink of an eye.

In his memoir *Bird of Passage*, Peierls claims that he and Frisch worked out their yield estimate "on the back of the proverbial envelope". This could not have been any ordinary letter-size envelope!

Appendix E
Equivalence of Peierls and Diffusion Theory Criticality Analyses in The Case of Large Critical Radius

The derivation in this appendix shows that the analysis of critical mass via diffusion theory presented in Appendix C is equivalent to Peierls' integral-equation approach in his 1939 paper in the case where the number of secondary neutrons is of order unity.

This Appendix shows how the results of the diffusion method of analyzing criticality developed in Appendix C and Peierls' integral-equation analysis give the same results when $v \sim 1$. This proof can be considered optional, but might appeal to more mathematically-inclined readers. The derivation is somewhat lengthy but is relatively straightforward.

The analysis developed here proceeds in three stages. First, Peierls' ξ parameter of Eq. (C.21) and his critical radius expression of Eq. (C.22) for $v \sim 1$ are re-cast to include the effect of scattering by introducing the scattering cross section σ_s. Second, the same is done for the diffusion theory, where we write down equivalents of various expressions in Appendix C which also include the effect of scattering. Third, the resulting main diffusion criticality equation, our modified (C.16), is considered for the case of a large critical radius. As described following Eq. (C.23), the dimensionless radial parameter $x = r/d$ of Eqs. (C.11) and (C.12) will be approximately equal to π when the radius is extremely large, that is, effectively infinite so that no neutrons escape. This results in an *overestimate* of the critical radius. This same approximation was invoked in subsection (iii) of Appendix D. To account for the finiteness of the core, a correction term δ is then introduced: $x = \pi - \delta$, where it is presumed that

δ is small compared to π. This is substituted into the criticality equation, and a series-expansion analysis is used to determine δ and hence x in terms of Peierls' ξ. This lets us ultimately write $r = xd$ in terms of ξ for comparison against Peierls' corresponding expression for the radius.

You would be well-advised to prepare some coffee or other stimulant.

To begin, the *total* cross section σ_t is defined as the sum of the fission and scattering cross sections:

$$\sigma_t = \sigma_s + \sigma_f. \tag{E.1}$$

σ_s would include the effects of both elastic and inelastic scattering, but the details of this need not concern us here. The corresponding total mean free path is then

$$\lambda_t = \frac{1}{n\sigma_t}. \tag{E.2}$$

Including scattering, Peierls' ξ of Eq. (C.21) becomes

$$\xi^2 = \frac{\sigma_f(\nu - 1)}{\sigma_s + \nu\sigma_f}. \tag{E.3}$$

In the denominator of this, set $\sigma_s = \sigma_t - \sigma_f$; the denominator becomes $\sigma_t + (\nu - 1)\sigma_f$, and an expression for $(\nu - 1)$ can be isolated; this will be useful in what follows:

$$(\nu - 1) = \frac{\sigma_t \xi^2}{\sigma_f(1 - \xi^2)} = \frac{\lambda_f \xi^2}{\lambda_t(1 - \xi^2)}. \tag{E.4}$$

Now, in the case where $\nu \to 1$, Eq. (E.3) indicates that $\xi \to 0$. In this regime, Peierls showed that the first line of Eq. (C.22) takes the analytic form

$$\frac{1}{\beta R_{co}} = \left(\frac{\sqrt{3}}{\pi}\right)\xi + 0.71\left(\frac{3}{\pi^2}\right)\xi^2, \tag{E.5}$$

where β is defined as the denominator of ξ^2 in Eq. (E.3) times a factor of the nuclear number density:

$$\beta = n(\sigma_s + \nu\sigma_f) = \left[\frac{1}{\lambda_t} + \frac{(\nu - 1)}{\lambda_f}\right] = \frac{1}{\lambda_t(1 - \xi^2)}. \tag{E.6}$$

In the last couple steps in this expression, β is cast in terms of ξ by first setting $\sigma_s = \sigma_t - \sigma_f$ as above, expressing cross sections in terms of mean free paths ($\lambda = 1/n\sigma$), and then eliminating $(\nu - 1)$ with Eq. (E.4). Exercise: Check that β has units of reciprocal distance.

The factor of 0.71 in (E.5) arises from Peierls' choice of boundary condition, that is, his version of Eq. (C.15); Serber had this factor as 2/3. The difference between

Appendix E: Equivalence of Peierls and Diffusion Theory Criticality Analyses ...

these choices traces to the fact that Serber assumed that the radius of the core is much larger than the transport mean free path so that he could treat the edge of the core as a plane, whereas Peierls adopted the results of an analysis which accounted for the curvature of the boundary. For sake of generality, I adopt the symbol f for this factor so that its propagation through the algebra can be tracked:

$$N_r(R_{co}) = -0.71\lambda_t \left(\frac{\partial N_r}{\partial r}\right)_{R_{co}} \rightarrow = -f\lambda_t \left(\frac{\partial N_r}{\partial r}\right)_{R_{co}}. \quad (E.7)$$

Note also here that it is the *total* mean free path λ_t that appears appears, as opposed to just the fission mean free path in (C.15); this is because scattering is now included.

Now define x as in Eq. (C.12),

$$x = \frac{r}{d}, \quad (E.8)$$

but with d of Eq. (C.11) defined to include scattering as

$$d = \sqrt{\frac{\lambda_f \lambda_t}{3(\nu - 1)}}, \quad (E.9)$$

where the α of (C.11) has been set to zero for threshold criticality; this will later let us write $r = R_{co}$.

Converting Eq. (E.7) to a derivative with respect to x gives

$$N_r(R_{co}) = -f\left(\frac{\lambda_t}{d}\right)\left(\frac{\partial N_r}{\partial x}\right)_{R_{co}}. \quad (E.10)$$

We now proceed to step two, casting the diffusion analysis in terms of Peierls' notation.

The convergent solution of the diffusion Eq. (C.2) modified to include scattering (in the last term, $\lambda_f \rightarrow \lambda_t$) is still the well-behaved part of (C.14),

$$N_r(r) = \left(\frac{\sin x}{x}\right), \quad (E.11)$$

where we drop the constant of integration A in (C.14) as it will play no role in the analysis.

On applying the boundary condition of Eq. (E.10), the analog of Eq. (C.16) is

$$x \cot(x) + \kappa x - 1 = 0, \quad (E.12)$$

where κ is a modified version of Eq. (C.17),

$$\kappa = \frac{d}{f\lambda_t} = \frac{1}{f}\sqrt{\frac{\lambda_f}{3\lambda_t(\nu-1)}}. \tag{E.13}$$

κ can be put in terms of Peierls' ξ parameter by eliminating $(\nu-1)$ via Eq. (E.4):

$$\kappa = \frac{1}{f}\sqrt{\frac{1-\xi^2}{3\xi^2}} = \frac{1}{\sqrt{3}f\xi}\sqrt{1-\xi^2}, \tag{E.14}$$

a result we will modify below and use later at Eq. (E.29).

Our task now is to "simplify" Eq. (E.12) for the case of $\xi \to 0$, although it will at first seem as if we make things more complicated. We take the terms in the equation one-by-one, beginning with κ in the middle term and using Eq. (E.14).

With the factor of ξ in its denominator, κ will clearly diverge for $\xi \to 0$. However, it turns out that κ gets multiplied by factors which converge more strongly that it diverges, so we run into no unphysical infinites.

We begin by performing a binomial expansion on the numerator of (E.14):

$$\kappa = \frac{1}{\sqrt{3}f\xi}\sqrt{1-\xi^2} = \frac{1}{\sqrt{3}f\xi}\left[1 - \frac{1}{2}\xi^2 - \frac{1}{8}\xi^4 - \frac{1}{16}\xi^6 \cdots\right]$$

$$= \frac{1}{\sqrt{3}f}\left[\frac{1}{\xi} - \frac{\xi}{2} - \frac{\xi^3}{8} - \frac{\xi^5}{16}\cdots\right]. \tag{E.15}$$

We will not need to go out to powers as high as ξ^5 for ultimately comparing with Eq. (E.5), but I will keep more terms than are needed through the algebra and truncate later on.

At this point we turn to the cotangent term in Eq. (E.12); we will return to κ later.

Now, as explained above, if the critical radius is large, then $x \sim \pi$ in Eqs. (E.11) and (E.12). We then write

$$x = \pi - \delta, \tag{E.16}$$

where δ is presumably small and to be determined.

A useful identity here is that $\cot(x) = \cot(\pi - \delta) = -\cot\delta$. Now invoke the Taylor-series expansion for the cotangent function:

$$\cot\delta = \frac{1}{\delta} - \frac{\delta}{3} - \frac{\delta^3}{45} - \frac{2\delta^5}{945} - \cdots. \tag{E.17}$$

This is valid for $\delta < \pi$. Again, this incorporates more terms than we will ultimately need to retain.

At this point, substitute Eqs. (E.16) and (E.17) into Eq. (E.12), keeping in mind the negative sign in the identity cited above. For the time being, κ is left as it is. After some algebra, you should find that (E.12) reduces to a series expansion in δ:

$$\pi = (\kappa\pi)\delta + \left(\frac{\pi}{3} - \kappa\right)\delta^2 - \left(\frac{1}{3}\right)\delta^3 + \left(\frac{\pi}{45}\right)\delta^4 - \left(\frac{1}{45}\right)\delta^5 - \cdots. \quad (E.18)$$

Now, our problem is to get a solution for δ to back-substitute into (E.16) to determine x and then get an expression for the radius in $r = xd$, but with the expansion in (E.18) this looks like an impossible task. Luckily, however, there is a little-known but extremely powerful mathematical technique known as "reversion of series" available to help us out.

The essence of reversion of series is that if you have a seres of the form

$$y = a_1 w + a_2 w^2 + a_3 w^3 + a_4 w^4 + \cdots, \quad (E.19)$$

where y and $a_1, a_2, \ldots$ are presumed-known numerical or analytic coefficients, then the variable w can be expressed as

$$w = A_1 y + A_2 y^2 + A_3 y^3 + A_4 y^4 + \cdots, \quad (E.20)$$

where the first few coefficients $A_1, A_2, \ldots$ depend on the $a_1, a_2, \ldots$ as

$$\begin{cases} A_1 = \frac{1}{a_1} \\ A_2 = -\frac{a_2}{a_1^3} \\ A_3 = \frac{1}{a_1^5}(2a_2^2 - a_1 a_3) \\ A_4 = \frac{1}{a_1^7}(5a_1 a_2 a_3 - a_1^2 a_4 - 5a_2^3). \end{cases} \quad (E.21)$$

A good mathematical physics textbook will give a more general treatment of this technique; in most problems one will need only the first few terms. What we are doing here is applying this technique to extracting an expression for δ in terms of powers of π and the coefficients in equation (E.18).

Comparing Eqs. (E.18) and (E.19) shows that we have we have $y = \pi$, $a_1 = \kappa\pi$, $a_2 = (\pi/3 - \kappa)$, $a_3 = -1/3$, $a_4 = \pi/45$, and $a_5 = -1/45$. To keep terms to κ^3, which will be more than enough for our purpose, we need only A_1, A_2, and A_3 in (E.20) and (E.21). These work out to

$$\begin{cases} A_1 = \frac{1}{\kappa\pi} \\ A_2 = \frac{(\kappa - \pi/3)}{\pi^3 \kappa^3} \\ A_3 = \frac{1}{\pi^5 \kappa^5}(2\pi^2/9 - \pi\kappa + 2\kappa^2). \end{cases} \quad (E.22)$$

Applying the recursion algorithm gives, after some algebra,

$$\delta = \frac{1}{\kappa} + \frac{1}{\pi \kappa^2} + \frac{(2 - \pi^2/3)}{\pi^2 \kappa^3} + \cdots . \tag{E.23}$$

Notice here that if κ diverges for an extremely large core, then $\delta \to 0$, exactly as we would expect.

A corollary result here is that we can get an explicit series expansion for the critical radius from Serber's formulation: $x = \pi - \delta$ and $R_c = dx = d(\pi - \delta)$, giving, with Eq. (E.9),

$$R_{co} = \pi \sqrt{\frac{\lambda_f \lambda_t}{3(\nu - 1)}} \left[1 - \frac{1}{\pi \kappa} - \frac{1}{\pi^2 \kappa^2} - \frac{(2 - \pi^2/3)}{\pi^3 \kappa^3} + \cdots \right]. \tag{E.24}$$

The inverse powers of κ here make this an expansion in terms of increasing powers of λ_t/d; see Eq. (E.13).

For the comparison with Peierls' result, we need $x = \pi - \delta$ by itself:

$$x = \pi \left[1 - \frac{1}{\pi \kappa} - \frac{1}{\pi^2 \kappa^2} - \frac{(2 - \pi^2/3)}{\pi^3 \kappa^3} + \cdots \right]. \tag{E.25}$$

To effect the comparison with Peierls, I write his $1/\beta R_{co}$ of Eq. (E.5) using Eq. (E.6) for β, set $R_c = dx$ from the Serber formulation, and express d in terms of ξ with Eq. (E.9):

$$\frac{1}{\beta R_{co}} = \frac{1}{\beta dx} = \frac{\sqrt{3}}{x} \xi \sqrt{1 - \xi^2} = \frac{\sqrt{3}}{x} \left[\xi - \frac{\xi^3}{2} - \frac{\xi^5}{8} - \frac{\xi^7}{16} \cdots \right], \tag{E.26}$$

where in the last term a binomial expansion of $\sqrt{1 - \xi^2}$ was carried out. Equation (E.24) for R_{co} could have been used directly here, but the following development for the factor of x in the denominator of (E.26) is a little more compact.

We need an expression for $1/x = 1/(\pi - \delta)$. Write this as

$$\frac{1}{x} = \frac{1}{\pi - \delta} = \frac{1}{\pi(1 - \delta/\pi)} = \frac{1}{\pi} \left[1 + \left(\frac{\delta}{\pi}\right) + \left(\frac{\delta}{\pi}\right)^2 + \left(\frac{\delta}{\pi}\right)^3 + \cdots \right], \tag{E.27}$$

where a binomial expansion was performed on $(1 - \delta/\pi)^{-1}$.

At this point the calculation becomes rather messy: Use Eq. (E.23) for δ, divide it by π, and substitute into Eq. (E.27). Keeping terms to powers of $1/\kappa^3$ gives, after some work,

$$\frac{1}{x} = \frac{1}{\pi} \left[1 + \frac{1}{\pi \kappa} + \frac{2}{\pi^2 \kappa^2} + \frac{(5 - \pi^2/3)}{\pi^3 \kappa^3} + \cdots \right]. \tag{E.28}$$

Appendix E: Equivalence of Peierls and Diffusion Theory Criticality Analyses ...

Now put κ in terms of Peierls' ξ parameter via Eq. (E.14). This gives

$$\frac{1}{x} = \frac{1}{\pi}\left[1 + \frac{f}{\pi}\left(\frac{3\xi^2}{1-\xi^2}\right)^{1/2} + \frac{2f^2}{\pi^2}\left(\frac{3\xi^2}{1-\xi^2}\right) + \right. \tag{E.29}$$

$$\left. + (5 - \pi^2/3)\frac{f^3}{\pi^3}\left(\frac{3\xi^2}{1-\xi^2}\right)^{3/2} + \cdots \right].$$

Expand out these terms, keeping contributions to only order ξ^2 for eventual comparison with Peierls' equation (E.5). After more algebra, you should find

$$\frac{1}{x} = \frac{1}{\pi}\left[1 + \left(\frac{\sqrt{3}f}{\pi}\right)\frac{\xi}{\sqrt{1-\xi^2}} + \left(\frac{6f^2}{\pi^2}\right)\frac{\xi^2}{1-\xi^2} + \cdots \right]. \tag{E.30}$$

Since ξ is small in this analysis, we can again do binomial expansions to second order:

$$\frac{1}{\sqrt{1-\xi^2}} = 1 + \frac{1}{2}\xi^2 + \cdots \tag{E.31}$$

and

$$\frac{1}{1-\xi^2} = 1 + \xi^2 + \cdots . \tag{E.32}$$

Using these in (E.30) gives

$$\frac{1}{x} = \frac{1}{\pi}\left[1 + \left(\frac{\sqrt{3}f}{\pi}\right)\xi(1 + \xi^2/2) + \left(\frac{6f^2}{\pi^2}\right)\xi^2(1+\xi^2) + \cdots \right], \tag{E.33}$$

or, on keeping terms only to ξ^2,

$$\frac{1}{x} = \frac{1}{\pi}\left[1 + \left(\frac{\sqrt{3}f}{\pi}\right)\xi + \left(\frac{6f^2}{\pi^2}\right)\xi^2 + \cdots \right]. \tag{E.34}$$

We are almost there. Put this result into Eq. (E.26):

$$\frac{1}{\beta R_{co}} = \frac{\sqrt{3}}{\pi}\left[\xi - \frac{\xi^3}{2} - \cdots\right]\left[1 + \left(\frac{\sqrt{3}f}{\pi}\right)\xi + \left(\frac{6f^2}{\pi^2}\right)\xi^2 + \cdots\right]$$

$$= \frac{\sqrt{3}}{\pi}\left[\xi + \frac{\sqrt{3}f}{\pi}\xi^2 + \left(\frac{6f^2}{\pi^2} - \frac{1}{2}\right)\xi^3 + \cdots\right] \tag{E.35}$$

$$= \frac{\sqrt{3}}{\pi}\xi + f\left(\frac{3}{\pi^2}\right)\xi^2 + \frac{\sqrt{3}}{\pi}\left(\frac{6f^2}{\pi^2} - \frac{1}{2}\right)\xi^3 + \cdots.$$

With $f = 0.71$, the first two terms in this last expression are exactly Peierls' equation (E.5), completing the proof! If you work back through the algebra, you should be able to convince yourself that this expression does correctly capture all the ξ^3 terms despite the truncation at Eq. (E.34). If you have followed all the way through this, congratulations! For fun, try working the calculation to order ξ^4.

Using modern parameter values for ^{235}U, how well does this analysis do? Chadwick (2021) gives the total cross section for ^{235}U as 5.0 barns; with $\sigma_f = 1.235$ barns as used in Chap. 4, we can infer a scattering cross-section of $\sigma_s = 3.765$ barns. He also gives $\nu = 2.57$. In Eq. (E.1), these give $\xi^2 = 0.2794$, or $\xi = 0.5286$. With $f = 0.71$, Eq. (E.35) out to the ξ^2 term then gives $1/\beta R_{co} = 0.3517$. With the number density of 4.792×10^{22} cm^{-3} from Sect. 4.5.1, Eq. (E.6) gives $\beta = 0.3325$ cm^{-1}. Hence we have $R_{co} = 1/[(0.3325 \text{ cm}^{-1})(0.3517)] = 8.55$ cm. At a density of 18.7 gr cm^{-3}, this gives a mass of about 49 kg, very close to the true value of about 46. That the agreement is this good is somewhat surprising as 8.5 centimeters is by no means "very large". (Question: How do these numbers change if you go out out the ξ^3 term in (E.35))? If you keep the first two terms in Eq. (E.15), you should find $\kappa \sim 1.3234$, and then the first two terms in Eq. (E.23) give $\delta \sim 0.9374$, which is by no means small compared to π as assumed in Eq. (E.16). However, as we saw in Appendix C, Peierls' expression for the critical radius for the cases of $\nu \gg 1$ and $\nu \sim 1$ do not differ wildly when the number of neutrons per fission is about 2. In a sense, the parameter values land in a sort of "sweet spot" such that the assumption of $\nu \sim 1$ gives a reasonable critical mass; sometimes one gets lucky.

Appendix F
Glossary of Symbols

The tables in this chapter summarize the meanings of the various symbols used in mathematical expressions in this book and give references to their defining equations (Tables F.1 and F.2).

Table F.1 Glossary of Symbols: Greek letters

Symbol	Name/description	Units	Primary equation(s); comments
α	Separation factor in criticality analysis	–	(C.6)
β	Factor in criticality analysis	–	(E.6)
γ	Pressure-energy density coefficient; factor in criticality analysis	– –	(D.9) (C.19), (C.20)
δ	Core radius correction factor; App. E	–	(E.16)
ϵ	Energy released per fission	MeV	(D.7)
η	Crtical radius/mean free path ratio	–	(C.18)
κ	Factor in criticality analysis	–	(E.12), (E.13)
λ_{deB}	De Broglie wavelength	cm, m	(4.9)
λ_f	Fission mean free path	cm, m	(C.1)
λ_t	Total mean free path	cm, m	(E.2)
ν	Neutrons per fission	–	Sect. 4.2
ξ	Factor in Peierls analysis of criticality	–	(C.21), (E.3)
ρ_o	"Normal" density of fissile material	gr cm^{-3}	
σ_f	Fission cross section	barns (bn)	(C.1)
σ_t	Total cross section	barns (bn)	(E.1)
τ_o, τ	Neutron population growth timescale	sec	Sect. 4.6, (C.4)

Table F.2 Glossary of Symbols: English letters

Symbol	Name/description	Units	Primary equation(s); comments
A	Atomic weight; constant of integration	gr mol^{-1}	(4.2), (C.14)
$c(t)$	Core expansion homology factor	s^{-1}	(D.10)
C	Integral of core expansion homology factor	–	(D.11)
d	Factor in criticality analysis	m	(C.11), (E.9)
D	Diffusion constant	m^2 s^{-1}	(C.5); $\lambda_f^2/3\tau_o$
DR	Decay rate	s^{-1}	(4.20)
E	General expression for bomb yield	Joules, kilotons	(4.17)
$E_{fiss}(t)$	Fission energy liberated to time t	Joules, MeV	(D.7)
E_{MeV}	Neutron kinetic energy	MeV	(4.5)
f	Numerical factor in Peierls criticality analysis	–	(E.7), (E.10)
k	Factor in criticality analysis	–	(C.16), (C.17)
M_{core}	Core mass	gr or kg	
n	Nuclear number density	nuclei per cubic m or cm	(4.2)
$\mathcal{N}(t)$	Total number of fission-liberated neutrons	–	(D.6)
N_A	Avogadro number	mol^{-1}	(4.2)
N_o	Initial neutron density	neutrons per m^{-3}	(C.8)
$N(r,t)$	Neutron number density	neutrons per m^{-3}	Appendix C
$N_r(r)$	Radial solution of diffusion equation	–	See $N(r,t)$
$N_t(t)$	Time solution of diffusion equation	–	See $N(r,t)$
$P(t)$	Pressure within bomb core, time t	Pascals	(D.9)
P_o	Factor in pressure within bomb core	Pascals	(D.9)
r	General radial position within core	m	
R	Core outer radius at any time	m or cm	
R_{co}	Critical radius for "normal" density ρ_o	m	(4.7), Appendix C
R_0	Initial core radius	m or cm	$R_0 > R_{co}$
R_{shut}	Core radius at criticality shutdown	m	(4.15)
t	General symbol for time	s	
t_{shut}	Criticality shutdown time	s	Appendix D
$U(t)$	Energy density, time t	Joule or MeV m^{-3}	(D.8)
$\langle v \rangle$	Average neutron speed	m s^{-1}	λ_f/τ_o; (4.5)
x	Dimensionless size factor in criticality analysis	–	$= r/d$; (C.12)
y	Yield of (α, n) reaction	–	Sect. 4.8

Appendix G
Physical Constants and Conversion Factors

The table in this chapter lists values for a few physical constants used in this book.

The National Institute of Standards and Technology (NIST) maintains an extensive listing of physical constants at physics.nist.gov/constants. The values here are taken from that list, truncated to four decimal places (Table G.1).

Table G.1 Physical constants

Quantity	Symbol	Value	Unit
Speed of light in vacuum	c	2.9979×10^8	m s^{-1}
Planck's constant	h	6.6261×10^{-34}	J-s
Planck's constant	$\hbar = h/2\pi$	1.0546×10^{-34}	J-s
Boltzmann's constant	k	1.3806×10^{-23}	J K^{-1}
Elementary charge	e	1.6022×10^{-19}	C
Neutron mass	m_n	1.6749×10^{-27}	kg
Avogadro's number	N_A	6.0221×10^{23}	mol^{-1}
MeV	Million electron-volts	1.6022×10^{-13}	J
kt	kiloton	4.2×10^{12}	J

References

Chadwick, M. (2021). Nuclear science for the Manhattan Project and comparison to today's ENDF data. *Nuclear Technology, 207*(S1), S24–S61.

Clark, R. W. (1965). *Tizard*. London: Methuen.

Lestone, J. P., Rosen, M. D., & Adsley, P. (2021). Comparison between historic nuclear explosion yield formulas. *Nuclear Technology, 207*(S1), S352–S355.

Pearson, J. M., & Reed, B. C. (2024). Remarks on the yield of fission bombs. *American Journal of Physics, 92*(9), 680–685.

If you have any concerns about our products,
you can contact us on
ProductSafety@springernature.com

In case Publisher is established outside the EU,
the EU authorized representative is:
**Springer Nature Customer Service Center GmbH
Europaplatz 3, 69115 Heidelberg, Germany**

Printed by Libri Plureos GmbH
in Hamburg, Germany